12.

SYSTEMS ANALYSIS IN BUSINESS

UNWIN PROFESSIONAL MANAGEMENT LIBRARY

1. COMPANY ORGANIZATION
   Theory and Practice
   PA MANAGEMENT CONSULTANTS LTD

2. MARKETING AND HIGHER MANAGEMENT
   ESMOND PEARCE,
   PA MANAGEMENT CONSULTANTS LTD

3. EFFECTIVE INDUSTRIAL SELLING
   with Total Marketing Communication
   VIC MARKHAM

4. MANAGING FOR PROFIT
   R. R. GILCHRIST

5. PLANNING THE CORPORATE REPUTATION
   VIC MARKHAM

6. MANAGING THE TRAINING FUNCTION
   using instructional technology and systems concepts
   CHRISTOPHER GANE

SYSTEMS ANALYSIS IN BUSINESS

JOHN GRAHAM

London
GEORGE ALLEN AND UNWIN LTD
RUKSIN HOUSE   MUSEUM STREET

First published in 1972

This book is copyright under the Berne Convention. All rights are reserved. Apart from any fair dealing for the purpose of private study, research, criticism or review, as permitted under the Copyright Act, 1956, no part of this publication may be reproduced, stored in a retrieval system, or transmitted, in any form or by any means, electronic, electrical, chemical, mechanical, optical, photocopying, recording or otherwise, without the prior permission of the copyright owner. Enquiries should be addressed to the publishers.

© John Graham 1972

ISBN 0 04 658134 0

Printed in Great Britain
in 10pt Times Roman type
by Alden & Mowbray Ltd
at the Alden Press, Oxford

Contents

1 THE WORK OF THE SYSTEMS ANALYST *page* 11
The Computer Age
Background to Systems Work
Structure of the Data Processing Department
Stages of Systems Development
Attributes of a Good Analyst
Scope of this Book

2 THE ASSIGNMENT BRIEF AND INITIAL
INVESTIGATION 30
The Right Beginning
Functional Relationships
The Assignment Brief
Why Investigate Current Systems?
Sources of Information
Fact Recording
Analysis and Appraisal of an Existing System
Summary of Investigation Objectives

3 THE FEASIBILITY STUDY 52
Why have a Feasibility Study?
How Benefits can be Evaluated
Assessing the Costs of a New System
The Dilemma of a Feasibility Study

4 THE APPROACH TO SYSTEMS DESIGN 61
First Review the Objectives
Specification of Output
Data Collection and File Processing
File Design
Systems Controls

5 DATA COLLECTION AND PREPARATION 74
The Importance of Data Collection
Punched Cards
Magnetic Tape Encoders
Paper Tape
Eliminating Redundant Operations

CONTENTS

    Character and Mark Recognition Systems
    On-line Input Devices
    The Approach to Data Collection

6 DESIGNING FILE SYSTEMS–MAGNETIC TAPE   93
    Basic Method of Handling Tape Files
    Basic Run Types
    Security of Data
    Batch Processing
    Input Programs
    Validation
    Editing
    Indexing
    Sorting
    Updating and Maintaining Main Files
    Reporting from Magnetic Tape Files
    Restarts
    Scheduling Implications
    Aiming for Efficient File Handling
    Timing File Processing Systems

7 DESIGNING FILE SYSTEMS – DIRECT ACCESS   126
    Principles of Direct Access Storage
    Types of Files
    Sequential Files and Indexing
    Random Files and Address Generation
    The Problem of Bucket Overflow
    Basic Timing Problems for Direct Access Files
    Timing Methods for File Processing

8 ON-LINE AND REAL TIME SYSTEMS   141
    The Nature of On-line Applications
    Time Sharing
    Multi-access Computing
    Reliability in On-line Systems
    Real Time Applications
    Fail Safe

9 CONTROL OF SYSTEMS DEVELOPMENT   147
    Stages of Systems Work
    Getting the Objectives Right
    Project Selection and Initial Investigation

Feasibility Studies
Designing the System in Detail
Planning for Systems Development
Control of Programming

10 DOCUMENTATION IN THE SYSTEMS
    DEPARTMENT   169
    The Need to Communicate
    Documentation Standards
    Feasibility Reports
    Working Papers
    The Systems Definition
    Operating Documentation
    Getting Control of Documentation

11 INSTALLING AND MONITORING A
    SYSTEM   183
    Programming Development and Testing
    File Creation and Validation
    Staff Education
    Clerical Procedure Preparation and Acceptance
    Changeover Plan
    System Monitoring and Evaluation

12 INTRODUCING CHANGE   194
    Sources of Resistance
    The User's View of the System
    Automation Phobia
    Fears of Higher Management
    User's View of the Systems Analysts
    Effects of Change upon Departmental Stability
    The Future Challenge

    BIBLIOGRAPHY   203

    INDEX   205

Chapter 1

The Work of the Systems Analyst

THE COMPUTER AGE

The second half of this century is characterized by the rapid acceleration of scientific and technical achievements. Many remarkable advances have been made; e.g. in nuclear physics, biochemistry, medicine, electronics, communications, as well as in industrial production and business organization. Some of these stand out more than others – man's first footsteps into space have caught the imagination of the public, potentially turning fiction into fact. Yet each achievement is dependent upon continued progress in other related fields; for instance advancement in the aerospace industry requires the concerted exploitation of success in many other fields.

The general-purpose digital computer is a tool which has a vital position in all these matters, and it assumes increasing importance. Yet computers are continually misunderstood by both the layman and the technologist who may have to use them. If economic progress is not to be hampered, the mystique of computers has to be removed, and we must all learn to use them in the same way that one uses a typewriter or an automobile.

Is this ideal really practicable? There is no doubt that computers are frequently misused, and to some extent the intricacies of the equipment have outstripped the potential that some organizations have to make use of it. The software specialist has done much to overcome the problem of communication with the machine, by producing compilers that permit users to write programs in languages relevant to their own training. For example, a language such as FORTRAN may be learnt fairly quickly, and is used by managers and technologists throughout the world from the central business planner to the petroleum engineer in the field.

Languages are not the answer to all problems. Computers have existed in commercial organizations since the early 1950s, and although rapid advances have been made in their application to business problems, many promised benefits remain unfulfilled. This is true particularly where one seeks to utilize

11

the computer's potential as a vehicle of communication – to store large volumes of data in a coordinated fashion so that one fact can be correlated with another on demand to satisfy the need for accurate, up-to-date information. Whenever one considers any application involving the collection of large volumes of data and the maintenance of data files in an integrated manner, the specialized skills of the systems analyst are needed. It is he who has to bridge the gap that invariably comes between the user and the computer. His talents are not vested purely in his skill as a data processing practitioner, but also in his overall knowledge of the business, of its data, and the relevance of information to good decision-making in the organization.

Computers have existed as a tool of man for a comparatively minute period of time; during our lifetime the systems analyst, the software developer, and the design engineer will probably change the state of the art considerably. It is to be hoped that such changes will make computers familiar to everyone – not as objects to be feared or avoided, but as valuable tools that remove much drudgery and free ourselves for more interesting and fruitful activities. Whatever the outcome, the systems analyst has a key role, both now and in the future, to use his specialized skills and organizing ability to achieve progress; in the final reckoning it is he who has responsibility to ensure that a system works.

Computers are extremely powerful and flexible tools. They are also expensive and complex and it must be clear by now that many computer projects have become financial headaches to management. Why is this so? Why is it that many managements have failed to reap the benefits that computerized systems afford?

First of all we have to acknowledge that general-purpose digital computers are ambivalent. Conceptually they can be used to aid any enterprise which needs to process and communicate information, but they have to be managed with insight into the technical and social problems which will arise.

At heart every good manager has some talent for systems analysis, and the intention in this book is to summarize the precepts and show how the computer has influenced the approach to systems design.

BACKGROUND TO SYSTEMS WORK

It would be useful at this point to give a concise answer to the question: What is a systems analyst? The first problem that we encounter, however, is that the term itself, although widely used, has many substitutes which seem to imply different functions: system designer, system investigator, analyst/programmer, systems and methods analyst, systems engineer, and so on. None of these expressions even mention the word *computer*, although many people immediately think of computer applications when considering these terms.

The functions performed by systems staff have in fact grown considerably with the development of digital computers in commerce and industry, but there are still strong relationships between the work of the systems department and the activities formerly associated with organization and methods and work study. Nowadays it is common to find these functions grouped under the control of a data processing manager. All are concerned with analysing the operations performed by people and machines, and improving working procedures and methods of communication to enhance the efficiency of an organization.

The development of computer applications in business organizations has tended to change the emphasis of the systems department. In the early 1950s the analyst with computer training was concerned largely with improving overhead costs by designing systems to operate in problem areas where large volumes of transactions had to be handled. Thus computing power was often confined to standard application areas, e.g. payroll, inventory control, production control and sales statistics. The systems analyst was sometimes regarded as a computer programmer capable of converting, say, a punched-card application into a computer system with greater power to handle larger data volumes at increased speed. The turnround of results was improved and, in many cases, so was the quality and variety of information available from the system. Appreciation of the potential information-handling qualities of computer systems has lead to a widening of the scope of systems departments.

The systems department has become an instrument for putting into effect major policies of the top management of organizations. Its place in the organization and its relationship with other departments is not based solely upon the data processing skills and techniques that it can bring to bear on problems, but upon ability to improve the communication of information to managers and to provide a sounder basis for decision-making and evaluating the effects of management decisions at every level.

It was not until the mid 1960s that educational establishments and computer authorities began to lay down a basic syllabus for training systems staff and to specify standard methods for tackling the various stages of systems work. It is to be expected that these standards will be greatly developed in the years ahead, but nevertheless, a substantial amount of practical information can be set down as the basis of any course on this subject.

The managers within an organization have a need to understand the process of systems analysis and particularly to familiarize themselves with the potential and problems associated with computers. The intention in this book is to provide the necessary background to prepare managers for a deeper involvement with data processing. The best results in systems design and implementation are invariably achieved where the management in the using department takes a firm interest and control in the project. The failures nearly always arise where management leaves the responsibility to the technical specialists.

It is against this background that the opening chapter of this book is written. An attempt is made to define the functions to be performed by a systems analyst, the environment in which he works and the personal qualities and technical ability required to do the job. At the end of this chapter it should be possible to appreciate the role of the systems analyst and understand in general outline the way in which his work has to be conducted.

STRUCTURE OF THE DATA PROCESSING DEPARTMENT

A data processing department may be considered as consisting of the following functional units:

   (i) Systems development;
   (ii) Programming;
   (iii) Computer operations.

Role of the Systems Department

The systems development department is normally responsible for developing clerical and computer procedures to meet the operational and information requirements of an organization. Systems are developed to meet specified aims of the company's management, and the basic stages in tackling a particular problem include:

   (a) Problem identification;
   (b) Analysis of current procedures;
   (c) Design and specification of proposed system including files, programs, and all data handling and data control procedures;
   (d) Documentation, presentation, and acceptance of proposed system;
   (e) Programming, program testing, and preparation of operating documentation;
   (f) System testing;
   (g) Education of user departments and computer operations staff;
   (h) Implementation of new system;
   (i) Monitoring operation of new system;
   (j) System review, if necessary.

These stages do not occur in distinct phases, but usually arise in the general sequence indicated above. The systems department is responsible for all these functions except possibly the programming function which may be done by a separate programming department; even so, the systems department will

THE WORK OF THE SYSTEMS ANALYST

have to prepare detailed written specifications of the programs required. The systems department is responsible for ensuring that the whole of a system works correctly as specified, and have to ensure that every participant knows his role and has the knowledge and resources necessary to perform his function when the system goes live. We shall discuss the functions listed above in more detail later, but first will look at other units in the data processing organization.

Role of the Programming Department

The programming department may be part of the overall systems development function as shown in Fig. 1.1, but may otherwise be part of computer

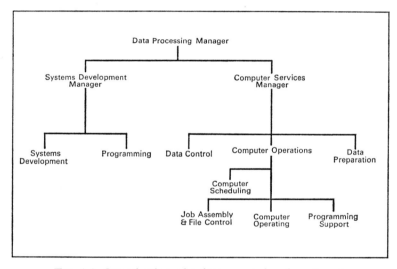

FIG. 1.1. Organization of a data processing department.

operations or even be a separate department in its own right. Programming is a specialized skill, and although most people can be trained to write programs, only experienced programmers are able to write, test, and debug programs efficiently so that they can perform the functions specified without errors. A programmer has to develop his programs in a logical manner, and has to chart and document the procedures so that other programmers can understand and amend them easily, if necessary, at a later date. The programming department usually has to work to a time schedule, and the final programs must be efficient in regard to the storage facilities and the computer time consumed when operational. As well as developing new programs, the programming department has also to support the programs currently running in

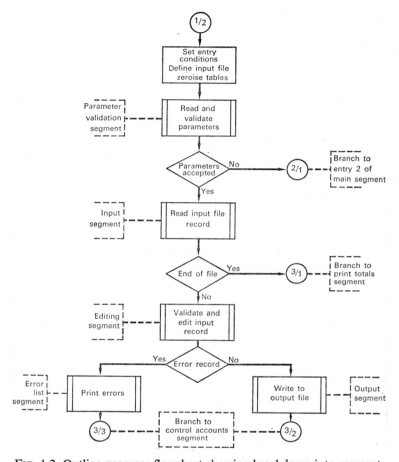

Fig. 1.2. Outline program flowchart showing breakdown into segments.

routine production – maintaining programs where necessary to meet system changes, or changes caused by external events, e.g. changes in tax rates.

The functions performed in programming include:

(a) Job breakdown;
(b) Flowcharting;
(c) Checking;
(d) Coding;
(e) Program preparation in computer input media;
(f) Compilation;
(g) Program testing and debugging;

(h) Documentation;
(i) Preparation of operating instructions;
(j) Program maintenance.

The job analysis is usually done by a senior programmer who studies program specifications given in the systems documentation provided by the

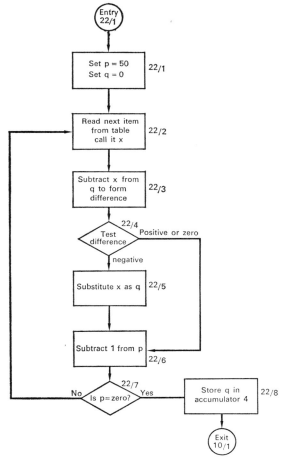

FIG. 1.3. Sample detailed flowchart showing a subroutine to select the largest of fifty positive values, stored in a table.

systems analyst; for each program he lays down the overall program strategy, perhaps drawing outline flowcharts showing the breakdown of the program into segments. If the program is large or complex, the separate segments may be produced by separate programmers. In any event, the next stage is the detailed flowcharting of each segment in which the operations to be performed are represented by logical symbols. The level of detail here is such that individual operations can be directly converted from this chart into coded instructions. Subroutines entered from a detailed flowchart are represented by single symbols cross-referencing to other flowcharts where the subroutines are themselves charted in detail. Samples of outline and detailed flowcharts are in Figs. 1.2 and 1.3.

The program is checked by dry-running through the detailed flowcharts using test data, and working with pencil and paper to simulate the operation of the program as though it were running live. This step highlights logical errors before the coding takes place.

Coding is the preparation of the source program in a computer language; this is a relatively simple task if the detailed flowcharting has been completed to the appropriate level. Coding includes narrative written on the instruction sheets alongside coded instructions to aid subsequent examination of the coded program. All coding sheets are numbered and cross-referenced to the flowcharts. A sample coding sheet is shown in Fig. 1.4.

Program preparation is done by the data preparation section (see Computer Operations below) who will punch the coded program into punched cards or paper tape. The program preparation is then verified by using data verification equipment.

It is unlikely that the original programs will be written in machine code, and therefore an appropriate language compiler will be used to check the validity of the coded statements and to convert them into the final machine language program (object program) on a specified input media. This operation, using the computer, will produce a list of program errors which offend the rules of the particular language. The errors have to be corrected and the program recompiled before continuing.

The program is then tested by the programmer using test data chosen by him to check every branch of the program. All output produced in this operation has to be carefully checked to see that the program has performed as desired. Logical errors in the program are thus detected and have to be corrected before testing the program with test data provided by the systems department. The systems analyst's test data must be chosen carefully to check all data conditions that can arise in live running of the system, and are usually prepared to test all programs in context of the whole system. The results of this test are returned to the systems analyst to check against known results prepared by him.

When program testing is completed, the programming department will finalize all program documentation and will file it in a standard manner so that it is available should subsequent maintenance of the program be necessary.

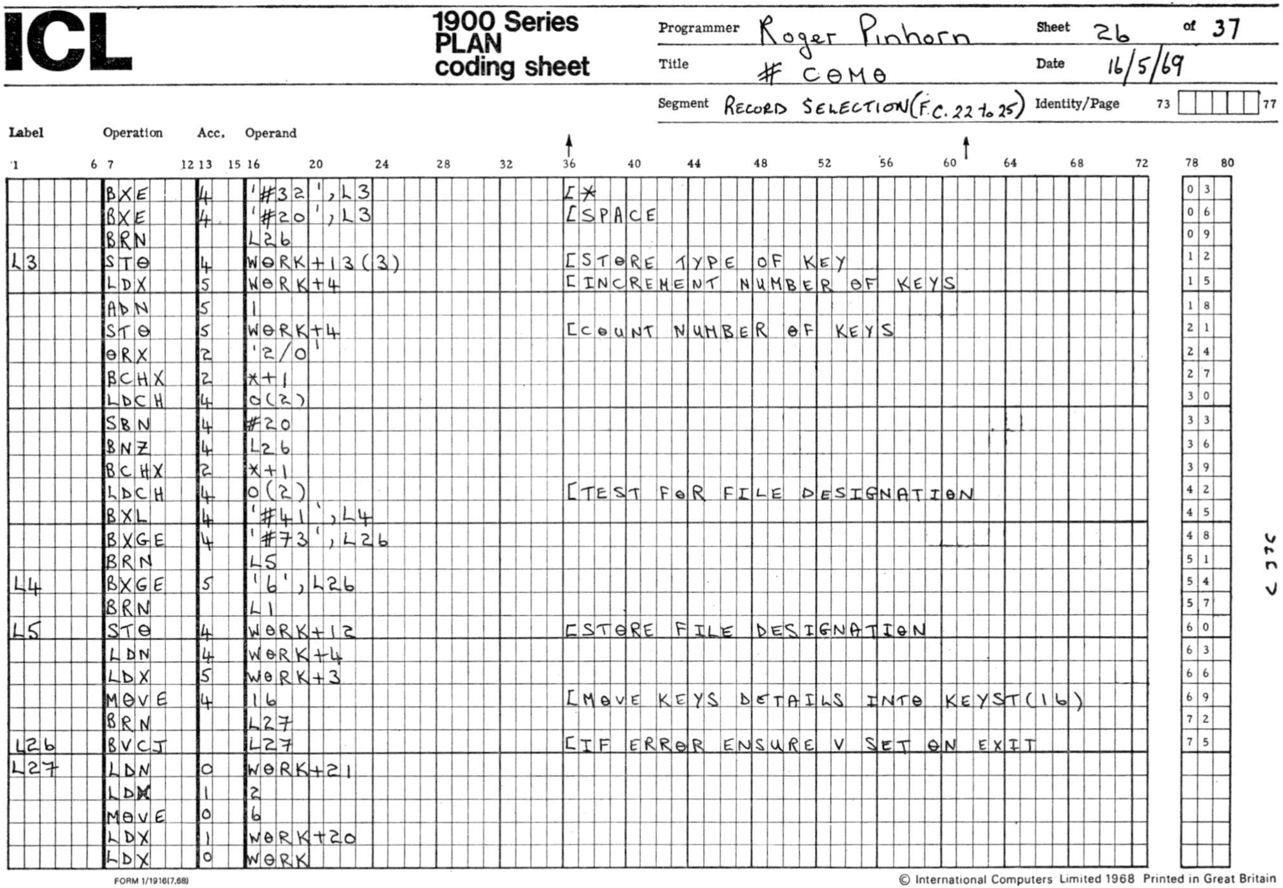

THE WORK OF THE SYSTEMS ANALYST

FIG. 1.4. Sample of program coding.

The programmers may then assist the systems analyst to prepare operating documentation for the system.

Role of the Computer Operations Department

The computer operations department has the task of processing work according to instructions given by systems analysts or programmers. Such work may be considered as:

(a) The processing of routine production work to agreed time schedules;
(b) The processing of *ad hoc* work; e.g. program testing or system development runs.

To process these jobs accurately and efficiently it is necessary that a number of other supporting activities be provided. The organization may differ from one company to another, but the following functions can be identified:

(1) Collection, vetting, and control of data;
(2) The preparation of data into computer input medium;
(3) The maintenance and control of the library of magnetic tapes or discs for live files, test files, program files, etc.;
(4) The assembly of all files and input/output media along with operating documentation for each job prior to its use in a live run;
(5) Operation of the job on the computer in accordance with written instructions provided for each job;
(6) Monitoring quality of results produced by the computer including checking accuracy of control accounts maintained by the system;
(7) Preparation of output for dispatch;
(8) Provide a support programming service to maintain the program library for minor changes conducted on behalf of the systems development department;
(9) Scheduling and control of the resources available in the operations department, and the maintenance of production schedules in co-ordination with user departments.

These functions are all shown in Fig. 1.1 as being under control of the computer operations manager; in practice this may not be so. For example, in large organizations, perhaps having a number of independent operating divisions, it is quite common to find that the control of data collection and preparation rests with the relevant line organizations. This work is conducted to standards developed and agreed jointly by the systems department and the line departments concerned, and is prepared on some input media to meet schedules for production of results. This practice has the advantage that user departments are encouraged to monitor and control the accuracy and timeliness of data arising within their own operation.

THE WORK OF THE SYSTEMS ANALYST

The support programming activity is often maintained within the systems development department, but where this is so it has to be accepted that effort is available to cope with critical problems that occur in current production work. For example, the recovery from erroneous data conditions or from magnetic tape failures sometimes requires special editing or report programs to be written quickly. At times like these, it is not merely the case that programming effort should be available, but that some knowledge of the data and the system itself should be on hand. The original systems development team may well be engaged in other areas or may have left the organization, and it is therefore important to preserve a knowledge of working systems within the operations team. This knowledge is, in any case, important in day-to-day co-ordination with user departments when work is operational.

STAGES OF SYSTEMS DEVELOPMENT

We have shown to date the basic functions performed by a systems department and the relationship of systems work to the other work done within a data processing installation. The functions performed by systems analysts are widespread. Initially, an analyst may be dealing with senior management in settling objectives and in reflecting the company policy in a systems development plan; later on, he is absorbed in the detailed design and specification of computer procedures, perhaps doing all the programming work and subsequently managing and controlling the education and implementation. These activities can be considered as falling within the following stages:

(1) *Problem identification* – setting major objectives and policy;
(2) *Feasibility study* – investigation of existing procedures and the development of outline plans to meet the desired objectives; the plans to be presented with estimated operational costs and the development costs and lead time;
(3) *Project approval* – presentation of plans for management appraisal and approval of terms of reference;
(4) *Detailed system design* – the development of detailed procedures and the specification of file layouts and programs;
(5) *Education* – the establishment and implementation of training plans to educate all those involved with launching and running the system;
(6) *Implementation* – the development and testing of programs and the production of all basic documentation for the system; the creation of data files, and the controlled introduction of new procedures;
(7) *Monitoring running system* – an audit and appraisal of the live procedures, with emphasis upon the accuracy of data in the system, and the effectiveness of the system in meeting its objectives.

SYSTEMS ANALYSIS IN BUSINESS

In most organizations, conditions will be changing rapidly and systems may well be under continuous development. Thus the life of a system may be represented as shown in Fig. 1.5.

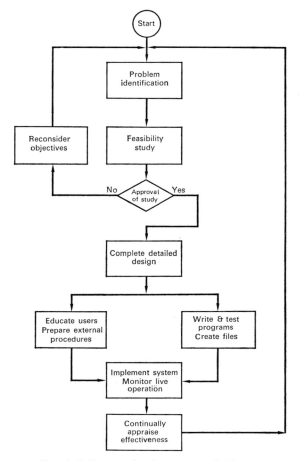

FIG. 1.5. System development activities.

ATTRIBUTES OF A GOOD ANALYST

Personality

A systems analyst has to be a person who quickly develops an awareness of the environment in which he is placed. He has certainly to be a good thinker, but not simply one who originates ten bright ideas a day. He is not a backroom boy or a boffin type, but preferably a person with a questioning mind

THE WORK OF THE SYSTEMS ANALYST

capable of understanding and appreciating the ideas of other people. He must be able to evaluate ideas not simply from the theoretical standpoint, but has to hold them up for examination against the present state of the organization in which he works.

It is no use producing an elegant solution to a problem if the organization is not yet able to benefit from such an approach. An idealist would probably be a danger within the systems department. However, the good systems analyst will always plan to achieve a near perfect solution in the long run, and will be able to present a plan taking perhaps several well-defined steps before achieving the ultimate goal.

He has to set a project on the right lines from the outset, and he must be capable of leading people and inspiring their confidence in the practicability of his ideas. Most major projects take several man-years to implement, and the credibility of the systems department would soon fade if the analysts were to go about creating false optimism. On the other hand, it is not unusual for user departments to get restless during the development of a long project: after the initial months the managers concerned will probably be looking forward to the satisfaction of some of their more pressing needs. An atmosphere of impatience is created and the systems analyst has to keep the project on the right rails while the detailed systems specification is developed and the computer programs are being written.

The ability to maintain confidence is therefore paramount, and it is necessary to review progress with the users in a sincere manner, helping to anticipate problems that do arise during implementation and to deal with those which threaten to interrupt the project. The mature systems analyst is a manager, and is responsible for many situations in which the lines of authority are not distinct. The systems analyst works in a constantly changing environment; he has to be ready and willing to accept change as part of his day-to-day scene, but has also to be able to apply pressure when and where necessary to protect his project from unnecessary or frivolous change. In this situation, his contacts and relationships with line managers and staff in other departments are all important.

Quite often, changes in the organization can have a major effect on a systems project; a new manager with fresh ideas and ambitions may so change the direction of an organization that much of the systems work has to be scrapped or drastically modified. It is at times like this when the systems man has to show his coolness and be able to pick up the pieces of his project once more. However, this type of occurrence should be rare: if the objectives of a project have been clearly established at the outset and have been accepted as part of a long-term plan for the organization, then changes should be more in the nature of modifications to the original plan.

The analyst needs a good sense of technical and political judgement and to be able to detect changes affecting the future success or usefulness of his project. Nothing is sadder than to observe the zealous systems analyst driving some moribund project towards a now irrelevant implementation date.

Whereas the systems man needs to convey his efficiency in organizing himself and members of his project team, he should not appear dogmatic. One essential personality trait for an analyst is the ability not to irritate people – he should never be arrogant, boastful, or in any way adopt a superior attitude towards his colleagues if he wishes to succeed. The business of project teams should be conducted openly and in such a manner as to make representatives of user departments feel that they want to help him make the project a success. In general, he should be generous about the contributions of others.

A good systems man should be receptive to new ideas and be able to create an atmosphere in which they can be allowed to develop. In his working life he must be prepared to suffer many irrelevant or incongruous suggestions without pain, and should be grateful if other people seize his best ideas and preach back to him.

The systems man has also to be a resourceful and energetic character, capable of working to a time-table and able to produce accurate and creative work under pressure.

So on and on the list of attributes grows until one wonders whether it would ever be possible to recruit anyone again with any confidence of success. But, of course, a systems department is an amalgam of people who have more or less the qualities described; the successful exploitation of these various talents depends much upon the ability of the systems manager and his senior project staff. Yet there still remains two essential qualities to be added to the list, and these are the possession of a mature attitude within the general business environment, and the ability to communicate effectively both with the written and the spoken word.

Communication is all important; it is not good enough just to have bright ideas – they must be put across in strength at the right time in the right manner. Every good solution has its pay off, and it is not immoral to be persuasive in getting good ideas accepted. Presentations and feasibility reports have got to appeal to management if any enthusiasm is to be maintained in development of the system. Documentation is also important in specifying computer programs, clerical routines, data collection procedures, and operating instructions, etc. Here the emphasis must be upon meticulous and well-organized presentation with a concise understanding of what the reader needs to know.

To complete this section, a check-list of the character traits considered desirable is provided. Whether we can always live up to them is another matter, but it does no harm to run through the list frequently to ensure that we give them a chance to work for us.

THE WORK OF THE SYSTEMS ANALYST

A creative thinker;
Aware of surroundings;
Practical outlook;
Planning ability;
Leadership and management ability;
Sincere and impartial;
Mature;
Responsible;
Patient;
Accurate;
Able to withstand and reconcile conflicting pressures;
Able to accept new ideas;
Able to sell good ideas;
A good communicator.

Technical Qualifications

A systems man must have practical experience of computer work, and, though he may not be required to write computer programs, should have had full experience in programming during his development. In some organizations analysts are expected to produce all their own programs, and in other cases will specify the details of programs needed so that the programming department can develop them. The type of ability necessary will differ from one company to another, but it is strongly recommended that at least one year's programming experience with an assembly language (and/or some high level language such as COBOL) should be attained.

The analyst has to think about the problems of the programmer in producing program specifications; the documentation created for this purpose must be complete and unambiguous, so that the programmer can carry out his work without having constantly to refer back to the analyst.

In many data processing units the programming department will only deal with major items of software, and individual analysts may have to write and test any one-off programs required in file creation to set up the system initially. As we shall see in future chapters, things do not always go smoothly in developing a system, and the systems man will have to sort out errors in programs and in the data itself. He will have to produce test data to test all the functional conditions of the system and will be ultimately responsible for getting the system to work correctly in a live environment.

The techniques and principles of file design must be mastered by the systems man, and these are often bound up with the hardware and software characteristics of the machines. A knowledge of these topics is therefore important if the power of the machine is to be efficiently utilized.

The staff in the systems department must keep abreast of new developments in the data processing industry if they desire to keep their organization as a front runner in its particular field. This does not mean that every new concept has to be introduced as soon as it appears. The data processing industry is somewhat notorious for selling ideas before they have been perfected; but new techniques have to be evaluated, and then, where appropriate, can be absorbed into the longer term plans of the organization.

It is the main purpose of this book to introduce many of these technical precepts, but the good systems analyst will never cease researching into new ideas. Many systems groups try to formalize this process by allowing individuals in the department to examine specific areas of the whole data processing field, each member being asked to update his colleagues with reports and oral presentations.

It is often said that someone entering the systems field needs to have a degree or an equivalent qualification. This is not essential in every case, but certainly a candidate needs to have the intellectual capacity to absorb the technical side of the business. Depending upon the nature of the work, it may be more important to have some commercial qualification in, say, accounting or business studies. There are, of course, more specialized appointments in the operations research field which demand, perhaps, a degree in computations or in one of the physical sciences.

One essential requirement for an analyst is a good knowledge of the industry in which he is located—with a particular awareness of that part of the organization in which he is working. If a systems designer is able to appreciate the history, development, and aims of the company, the more he will be aware of his own role and will see how he can contribute to the corporate objectives.

The data processing field is becoming increasingly specialized, but the systems analyst has to be an adaptable fellow and therefore needs to have a broad technical knowledge. I have already stated that the analyst should have some general programming experience; the implications are that he should understand computer equipment including the detailed functions and relationships of the basic hardware units and software that comprise a computer system. Practical knowledge of computer languages for both commercial and scientific purposes is very valuable.

The extent of the software knowledge may depend upon the immediate environment, but as a minimum it is advised that the utility and diagnostic software available for the particular computer should be understood as well as any relevant software applicable to the application area in which he is working. Above all, the analyst should understand the documentation and organization of the software library relevant to the equipment with which he works, so that he is able to take advantage of all general purpose software, both in system design and in tactical situations that arise in file creation and systems implementation.

THE WORK OF THE SYSTEMS ANALYST

Summary of Technical Requirements

The knowledge required by an experienced systems analyst can be summarized under the following headings:

(1) *Information requirements* – a knowledge of the industry in which the analyst operates, the organization and structure of the company, how management control is exercised, and the nature of the information needed by management in the company to make effective decisions;

(2) *Data processing* – an awareness of the role of the data processing department, the functions that it undertakes, and the way in which control can be exercised over the deployment of its resources to ensure that projects are correctly costed and evaluated. This will include an appreciation of the importance of each stage of systems development (project identification, investigations, feasibility studies, detailed systems design etc.);

(3) *Human relations* – the ability to understand the problems that new systems create for employees, customers, management, etc. The need for consultation and the necessity for agreeing and planning changes. The development of training schemes and the effects of new systems upon industrial relations generally;

(4) *Programming and computing* – a thorough appreciation of the organization of hardware units and software in a computer system. A working knowledge of one or more programming languages, and of the library of programs available for use with the equipment, working experience of programming and programming techniques; plus understanding of the implications of multi-programming, time-sharing facilities, and of operating systems and supervisory programs generally;

(5) *Data control and handling methods* – an awareness of the nature of information, of ways in which data can be captured and recorded, coding techniques, and methods for transmitting and receiving data. One point to emphasize is the need to understand the principles of data security as it applies at every stage in a system;

(6) *File processing and maintenance* – the organization of data in backing storage, principles of file and record design. The use of different storage devices and their relevance to general systems applications. File maintenance techniques and methods of reporting for management control and data control. An appreciation of operational conditions in a computer installation and the relevance of such conditions to program design and file processing generally;

(7) *Principles of computations and applications* – under this heading it is possible to consider background subjects in mathematics and statistics; e.g. numerical calculations, numerical errors, matrices, representation

of functions, sampling, standard deviation, probability theory, linear programming, and so on. It is necessary to consider these in relation to the particular business environment and the basic applications undertaken; e.g. inventory management, modelling, and so on;

(8) *Work measurement and work study* – this is often considered as a specialized branch of systems work not necessarily associated with computer applications. It entails the study and analysis of operations performed by qualified individuals in carrying out specified tasks in factories and offices. Each task is broken down into its component parts to determine the best sequence and method for completing the operations. The individual elements of the job may be timed and used to form an assessment of the work content of the job. Allowances can be included for relaxation periods and for other factors – such as enforced idleness due to uneven work flow. The results obtained can provide a proper basis for scheduling, planning, and controlling work. Many of the clerical functions which support computer systems need to be subject to this type of study, e.g. in order to ensure that adequate resources are available for data control.

SCOPE OF THIS BOOK

Many of the topics mentioned above are separate disciplines in their own right and as such can only be dealt with briefly in this work. The main aim of this book is to introduce and describe the particular skills that any experienced systems analyst must acquire through training and experience, and to suggest reading that will assist in the further development of each subject. The following subjects, although mentioned in some chapters, are not treated in any depth in this book:

Principles of programming;
Computations and applications;
Work measurement and methods study.

Anyone approaching this book without general grounding in programming will find certain sections hard going, but otherwise need not be concerned with these subjects. There are many publications which deal with the principles of hardware and software, and some are listed at the end of this book.

The organization of this book is as follows. The initiation of projects and the approach to understanding and establishing the nature and bounds of a particular assignment are described in Chapter 2. Methods of collecting information and analysing the related procedures that operate within an organization are also dealt with. Chapter 3 continues with an appraisal of the uses and methods of conducting a feasibility study and suggests methods of

THE WORK OF THE SYSTEMS ANALYST

establishing specific goals for a new project and evaluating its costs and benefits in outline.

Chapter 4 introduces the overall approach to the design phase of systems work, and is a preliminary to the next three chapters in which some specific guidelines are given for designing data collection and file processing procedures. Basic file processing techniques are given for magnetic tape files (Chapter 6) and direct access files (Chapter 7).

An introduction to the characteristics of on-line systems is given in Chapter 8; this is intended to demonstrate some of the complexities associated with this type of work, and further reading is recommended for more advanced students.

Chapters 9 and 10 relate to the management of a systems department and are important in describing the general environment in which the analyst works. The procedures encountered in particular data processing departments will vary, but the standards and control methods outlined in these chapters are widely and successfully employed in the initiation, development, and control of systems work.

Chapter 11 returns to the latter stages of a systems project and lays down the essential ingredients of an implementation plan. It should help the student analyst to appreciate the difficulties of installing a project and enable him to construct a plan that will aid in bringing his design work to fulfilment as a running system.

Finally, in Chapter 12, the social attitudes of all those concerned with a project are brought into focus. It is impossible to lay down hard and fast rules that will guarantee acceptance and commitment to a new system, but consideration of the views expressed in this chapter will materially improve the chances of installing a worthwhile system.

Chapter 2

The Assignment Brief and Initial Investigation

THE RIGHT BEGINNING

Any systems study should be conducted to ignore existing conditions or organization structures that might otherwise serve to curtail a creative approach to the organizations problems. This is particularly true when a company first buys a computer, because the quality of the work done in the beginning has far-reaching effects on the subsequent efficiency of the organization.

Difficulties often arise because many computer projects are conceived to attack particular application problems which have been identified as sore spots in the operation of the enterprise. Production control and inventory control are typical examples of application problems that attract attention in this manner.

The traditional method of tackling an application of this sort is to start by investigation and documentation of the existing procedures and departments concerned, looking for weaknesses in the flow of information and the control of operations. This approach is justified if the organization, with due consideration, is prepared to dedicate its data processing resources to a particular application. However, far greater potential rewards are forthcoming if systems can be conceived to embrace all the major functional aspects of the whole organization.

One hears reference to integrated system planning whenever systems people meet together, and the term holds a certain mystery for the newcomer to the data processing field. In fact, designing systems of this nature is not particularly difficult or different from the traditional application oriented approach, but it does entail a more perceptive study of an organization and its functions at the outset. The important phase is the identification of the bounds of each project and an understanding of its relationship to the adjacent functions within the organization.

The establishment of the terms of reference for a new project is a vital step before looking in any detail at existing procedures.

THE ASSIGNMENT BRIEF

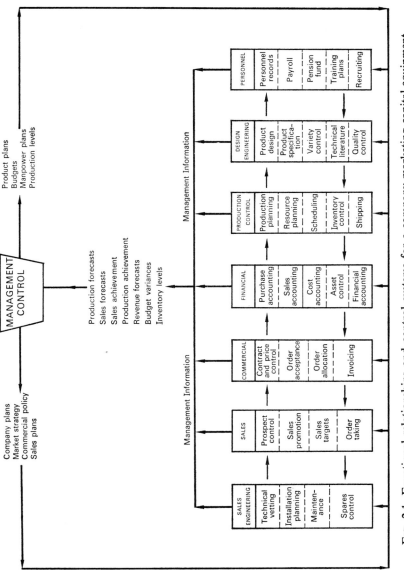

Fig. 2.1. Functional relationships and control systems for a company marketing capital equipment.

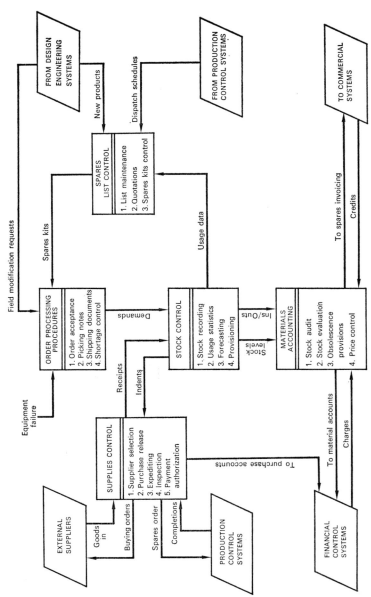

Fig. 2.2. System areas and functions in spares inventory control.

THE ASSIGNMENT BRIEF

FUNCTIONAL RELATIONSHIPS

The terms of reference should identify the major functions to be covered by a project and specify its areas of interface with existing systems and systems under development. Figure 2.1 shows the major functional relationships that might exist in a company manufacturing and marketing capital equipment for use in industrial organizations. It is not conceivable that all these functions could be handled as one systems project, but instead a number of subsystems would be planned to embrace major areas and to cover subfunctions within those areas.

In an organization of this nature there might well be a number of fairly large and relatively autonomous divisions. The structure of these divisions should represent the functional relationships of the various units, but it must be accepted that in practice such distinctions are not always clearly observed. The systems department can best achieve an overall plan by identifying these functions and ignoring, for the time being, any organizational constraints. Fig. 2.2 depicts the functional relationships that can exist in an organization established to control a spares inventory. Such a unit might well be part of the sales engineering support function shown previously in Fig. 2.1. The functions identified at this lower level represent in themselves major problem areas, and as we can see in Fig. 2.2, warrant implementation as a number of separate but closely related tasks.

To be able to observe all the functions correctly and to perceive their relationships will probably entail a very detailed initial study of the organization, its products, its aims, and objectives. This itself implies close cooperation with senior management throughout the group, and support by the directorate in starting and conducting the whole project. It is through such studies that the management will be able to observe weaknesses in the organization and will perhaps restructure the various units to reflect the functional relationships prevailing. This whole activity will not normally imply a detailed investigation of the procedures and documentation – this aspect will arise later when subsystems are tackled.

It need not follow that reorganization will be necessary. All systems must expect to operate across departmental boundaries at one stage or another. The systems staff have to be careful to see that such matters are discussed at the appropriate management level, and that the related social problems are not allowed to create resistance to the project as a whole. The approach to this aspect of the work is dealt with later in Chapter 12.

Fig. 2.3 shows the *before* and *after* condition that could occur in the inventory control department mentioned above. The changes created here are exaggerated for the benefit of the example, but it does help to show how

SYSTEMS ANALYSIS IN BUSINESS

necessary it is for the analyst to be able to appreciate the functional relationships and to ignore artificial relationships portrayed by some organization charts.

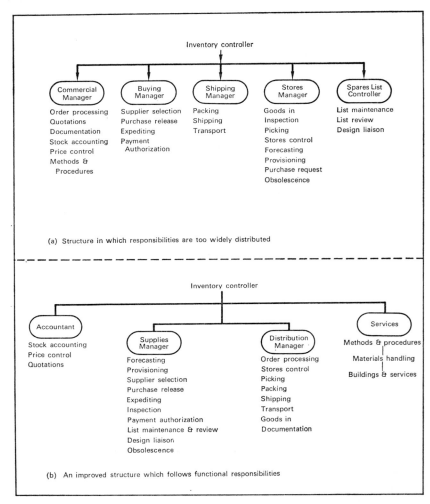

FIG. 2.3. Examples of organization structure in an inventory control unit.

Advantages of the Functional Approach

The functional approach is relevant to any project since all systems can be viewed as containing further subsystems or as being part of another system. The real advantage stems from the fact that the whole can be visualized as

THE ASSIGNMENT BRIEF

containing a number of bricks which can be put together as necessary to meet the overall objectives. Management are thus able to visualize the future shape of their control system, to know how far work has progressed, to know what the capability is in the months or years ahead, and to plan and budget for the resources needed to make further progress.

To the individual systems analyst it helps him to view his project within the overall framework, and the systems manager is able to identify the salient features necessary to ensure adequate co-ordination and integration of development work. In Chapter 9 the methods for controlling systems activity are developed in more detail from the theme described above.

THE ASSIGNMENT BRIEF

An assignment brief is a document which specifies and authorizes a systems project. The nature of the assignment brief is related to the size of the problem to be tackled, but it should be sufficiently concise to be read by senior executives who may be asked to approve it, and yet present the systems analyst with sufficient indication of the functions to be covered. A simple assignment is shown in the example in Fig. 2.4. If the problem is complex or likely to entail heavy expenditure for new resources, the assignment brief may call for deeper investigation to establish the feasibility of achieving the objectives. However, if the problem is relatively straightforward and falls within the current capacity of the data processing department, the assignment brief may request that detailed design work be started immediately and a systems definition be produced. In either case it is unlikely that programming will commence directly unless the project is, say, a simple reporting application from an existing computer file.

The work of completing the feasibility report or of doing any detailed system design will generally be proceeded by an initial investigation into the current operating procedures related to the functions under review.

WHY INVESTIGATE CURRENT SYSTEMS?

The assignment brief will usually provide a succinct statement of the nature and bounds of a project before a detailed investigation is carried out although, sometimes, a preliminary investigation is made in order to identify the problem more closely before the assignment brief is written. The purpose of a detailed investigation is to create necessary background information just before system design or feasibility studies are attempted. Although it may be relatively straightforward to establish the major functions in a particular system area, it is far more difficult to recognize the many events that might be necessary to the operations involved. An appreciation of side effects on other systems is also important at the time of starting to develop functions in detail.

If the analyst has little experience of the problem area he may benefit from the investigation in that he is able to study the many procedures necessary to operate the current system, but he should take care not to be unduly influenced

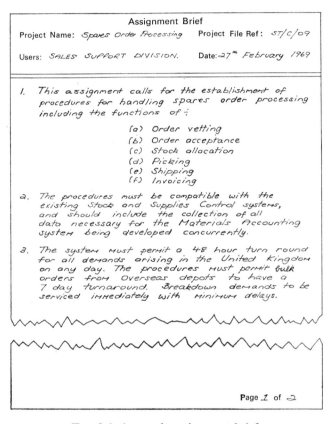

FIG. 2.4. A sample assignment brief.

by the procedures and organizational idiosyncrasies that may be uncovered. Too often opportunities for improving system efficiency, or for reducing operational costs, are thrown away because of a reluctance to change aspects of a previous system. Sometimes clerical routines are left to operate much as before save that data are collected at certain points to be fed into a computer file; this approach is not to be recommended because it leads to a lack of involvement and motivation in user groups.

THE ASSIGNMENT BRIEF

An advantage of such an investigation is that it does reveal the qualities and character of the individuals who run the current system. Most analysts heave a sigh of relief when they encounter a strong personality at supervisory level in a department under study. Even though such a person may initially react to the investigation, once his confidence has been gained he will be invaluable in identifying sources of potential trouble, and, later on, will help to get across the ideas necessary for the implementation of the project.

The terminology used within the problem area is also important – in many cases, business terms take on a particularly local flavour, and the analyst will need to communicate with staff in a manner that they understand.

The most important aspect of a detailed investigation is to discover all the occurrences that are liable to upset the operation of a scheme – the analyst has to be very attentive to statements like 'in 99 per cent of the cases the pink copy of the form is sent to the chief storekeeper'. It is the exceptional occurrences which are likely to undermine a new system unless they are identified and catered for in the new procedures. The important thing is to identify exceptions, measure their occurrence in practice, and design exception routines to deal with them.

One has to admit that there is a certain flair in knowing how far to go with an investigation. Many analysts feel that there is too much emphasis on detailed procedure analysis in earlier teachings about systems work. It is far less costly in time and effort to identify the main functions and the relevant people responsible for them and try to design procedures which meet these functions simply and effectively. Certainly it is not worthwhile constructing a vast volume of written material about an existing system–in many instances changes are going on perpetually and the analyst ends up chasing his tail to record all the details. An investigation is needed to ascertain:

(a) The functions needed;
(b) The functions performed;
(c) An evaluation of the actual performance;
(d) Who performs the functions;
(e) What information is needed;
(f) What information is currently available;
(g) How is information communicated;
(h) What data elements are required;
(i) What are the characteristics of the data elements.

An investigation is useful because, apart from the factors given above, it allows the analyst an opportunity to establish contact with the people in user departments and helps him to identify the strengths and weaknesses of the current operation. Above all, it should enable the analyst to ascertain the performance of the present system in relation to the expectations of management.

SYSTEMS ANALYSIS IN BUSINESS

The rest of this chapter deals with methods of fact recording, analysis of facts, and the development of fruitful contacts with user staff.

SOURCES OF INFORMATION

Information about an existing system can be derived from observation of the actual methods and procedures in use, by studying existing procedure manuals and systems definitions, and by questioning people involved in the operation of the system.

At the beginning it is a good idea to set up an investigation file into which copies of documents, reports, and interview notes can be placed as the investigation proceeds. The creation of such a file helps to organize the investigation; certainly the writing of notes and reports helps to fix information in the mind of the analyst, but in practice one finds that the file is rarely used for reference after the initial stages. It is probably best on balance to fix a definite period for completing the investigation and the creation of the investigation file, and try to cover the ground thoroughly in the scheduled time.

Existing Documentation

The collection of written material is a useful start, and among the most helpful documents are organization charts, procedure manuals, job specifications, departmental terms of reference, administrative instructions, commercial policy statements. All such material should be considered as background information and must be checked with the appropriate authority before placing too much reliance upon it.

When this existing material has been studied it should be considered in relation to the functions identified for the departments concerned. This may immediately raise a number of questions in the mind of the analyst which will form the basis of the next stage of the investigation.

For example, one may find that functions appearing in the terms of reference of a particular unit are not reflected in the job specifications of members of the department or in the organization charts available. Very often the reverse is true – the latest organization charts show that functions are being tackled which are not reflected in procedure manuals, and therefore someone somewhere is fulfilling a function not clearly identified in his terms of reference.

It is probable that any such documents as are available are contradictory simply because they have been produced at different times in the life of the organization. If the analyst is lucky he may find himself working in a stable atmosphere where it has been possible to keep such documentation up to date.

Questions and Interviews

Once the initial study of documentation has been completed, the analyst should draw up a plan to interview relevant members in the organization. Interviews must be carefully prepared. Steps should be taken to acquaint people with the topics to be discussed beforehand, and to keep them informed of the objectives of the whole study. The analyst should always try to give several days notice of impending interviews, and should seek permission of managers before approaching their staff. A little old-fashioned courtesy is very helpful in creating the right atmosphere. Whenever possible, interviews should be arranged in a quiet office where distractions can be avoided.

The person interviewed should not be left without a clear understanding of the reasons for the discussion. As far as possible, interviews should be of a conversational nature, and not question and answer sessions in which the analyst is continually taking notes. The person being interviewed should be allowed to do most of the talking. The analyst should state that he intends to produce notes, and ensure that the person concerned is allowed to read them subsequently in order to verify and amend any details necessary.

The interviews should be constructed to form an overall pattern of facts which can serve to check statements made by other parties. There is nothing sinister in this approach, it is simply that people tend to be inaccurate about the activities of others in an organization. Managers are liable to be misinformed about many of the detailed steps that their staff perform, but they are aware of the functions and general policies operating within their departments. Local methods and *ad hoc* or informal procedures are often adopted by people at the operating level to keep a system running smoothly.

The main aim of the interviewer is to keep going in the required direction, but he should avoid leading the person by asking questions which are too direct. Certainly avoid framing questions in such a way that monosyllabic answers are encouraged. It is often a good idea to appear a little confused by the detail so that the same ground has to be covered in a slightly different direction later on.

Group Sessions

In making an investigation of a large procedure in which many documents and forms are involved, it is often a good idea to run project teams to look into the details of particular procedures. People who handle forms or procedures can be brought together round a table to record the latest statement of the actions involved. It is often surprising to realize the extent to which people misunderstand the activities of others; often copies of forms and documents are still circulated and completed long after some aspects of the procedure concerned have become redundant.

On the other hand, one also finds procedures stretched to their limits with additional data elements being entered at odd places on various copies of the forms, and perhaps extra photo-copies being made to keep other branches informed of certain events.

Project teams set up in this manner must be controlled to cover specific topics in as few meetings as necessary. All members of the team should be given the opportunity to specify their involvement, and afterwards should receive written summaries of the sessions for verification and approval. All findings must be subsequently verified by observation of live events.

Observation of Live Procedures

All information gathered has to be vetted to see that a correct understanding of the current position is obtained. It is important to conduct the investigation at every level within a department to make sure that the system is examined from different view points.

All documentation should be studied in the live environment; blank copies of form sets and record cards are not particularly useful. For that matter, completed documents may be misleading if they are derived from archives. The best way of vetting the methods in current use is to spend a day or so working at key points as an operative in the system. In any event it is desirable to review live batches of documents each day as the investigation proceeds.

Often information is conveyed by more informal means: by telephone, memos, hand-written notes, etc. These are likely to be vital links in the system and should not be ignored.

FACT RECORDING

An investigation file should be opened at the outset; the form and structure of this file should be in line with standards within the department. A proposed structure and content for such a file is described below.

Section 1 – Problem identification. This section should contain a copy of any assignment brief that has been produced, or of correspondence or notes which serve to indicate the functions to be investigated and reviewed.

Section 2 – Organization. Existing organization charts may be included here, but it is more informative to redraw charts to follow the general form of that shown in Fig. 2.5. In this sample the chart identifies the major functions performed, the reporting structure, and the number of staff in each section identified by grade.

A separate list should accompany the chart giving the names of people concerned, their grade, sex, and telephone number. This list will serve as a handy reference for contacts in the subsequent investigation of the system. Within this section, administrative or organizational instructions giving terms of reference for the departments should be included. If such documents are not

THE ASSIGNMENT BRIEF

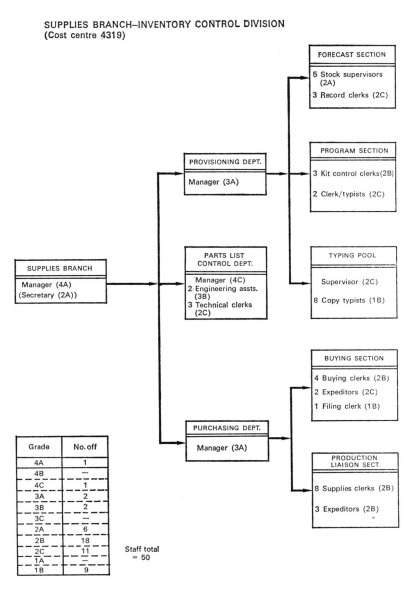

Fig. 2.5. Organization chart showing function and grades of staff.

available then the analyst should ask the relevant managers within the organization to define their responsibilities and should commit descriptive notes to the file.

FORM IDENTIFICATION	FORM DESCRIPTION - Page 1 of 1
Name: Stores Demand Note Form No: LM 171/9 (19·3·67) Procedure: Spares Issuing Procedure	
PURPOSE To permit Maintenance staff to order Materials from the Central Spares Store, and to collect all data necessary for the spares order processing routine.	VOLUMES 600 notes of average 5 items per note per day.
DISTRIBUTION Copy 1: To Order Dept. in Central Spares Store. Copy 2: Retained by engineer until parts are received, then sent to Accounts Section.	

DATA ELEMENTS			
Item	Format	Size	Origin
Spares List No. & Page	Alphanumeric fixed	A 9999/99	Completed by Maintenance engineer when ordering parts.
Line No.	Numeric fixed	99	
Quantity	Numeric variable	999999 max.	
Equipment Serial No.	Numeric fixed	999999	
Engineer's No.	Numeric fixed	9999	
Site Address Code	Numeric fixed	99999	
Form No.	Numeric fixed	999999	Pre-Printed

REMARKS This form is a satisfactory punching document, but Form Numbers are not unique at present, and there is no recognised space to enter the date.

FIG. 2.6. Sample form description sheet.

Section 3 – Procedural descriptions. This section may include or refer to existing documents, specifications, and procedure manuals along with notes outlining any inconsistencies identified as the investigation proceeds. Flowcharts and supporting narrative may be produced if it is helpful, and all major procedures, documents, and records should be identified.

Section 4 – Form descriptions. A sample of every form used should be filed along with narrative that describes the number of copies used, the data elements entered, who enters what data elements, who uses what data elements,

FILE IDENTIFICATION		FILE DESCRIPTION - Page *1* of *1*	
File Name:	Master Spares List.		
Freq. of Updating:	Weekly.	Media	Punched Cards
PURPOSE To provide a master file for printing copies of spares lists, for assembly of kit orders for Central Spares Depot, and for issuing quotations.		VOLUME 750 separate lists of average 3,000 entries each.	
SEQUENCE List No. by Page by Line No.			
DATA ELEMENTS Item	Format	Size	Notes Columns
List Number	Alphanumeric Fixed	A9999	1 to 5
Page Number	Numeric fixed	99	6 to 7
Line number	Numeric fixed	99	8 to 9
Description	Alphanumeric fixed	20 characters	10 to 29 (left justified)
Supplier ref.	Numeric fixed	999999	30 to 35
Supplier	Alphanumeric fixed	15 characters	36 to 50 (left justified)
Quantity	Numeric variable	999 max.	51 to 53
Page suffix	Numeric	9	54 see notes below
REMARKS A useful source for the new master file. Note, descriptions are not standardised and need extension. In one or two cases, lists have been generated which needed more than 99 pages.			

FIG. 2.7. Sample file description sheet.

when and how they are used. A simple form can be used to record this data on the lines suggested in Fig. 2.6. The size and structure of data elements should be ascertained and included in this list as shown. This last point is vital if one is considering the use of mechanized files or computer records for recording transactions and summaries of the data. It is better to allow for worst case conditions rather than underestimate the values that fields and variables may take. Volumes of documents or items handled daily or weekly, etc., should be given.

Section 5 – File formats and contents. Files may exist as manual records, as punched cards, paper tape, or magnetic tape, and so on. Obviously records which exist in some mechanized form are very useful and important in setting up a new system, and it is likely that up-to-date specifications may already exist. However, manual records can quite often be overlooked, and the analyst should take particular care to ascertain the contents of such files and uses made of them. The size and format of data elements and the volume of records have to be defined. A method of recording details of files and records is given in Fig. 2.7.

Section 6 – Interview notes and project team minutes. The details collected by questioning and observing current activities have to be recorded in an orderly manner. This can be left to the discretion of the analyst according to the type of problem being reviewed. The most obvious choice is to file papers by function, but often several functions are served by a particular form, and therefore it may be better to arrange documents by the procedures involved. In any case, a simple contents list or index should be provided.

Interview notes and project team minutes may need to be rewritten to provide reports and summaries for future reference. To minimize this sort of activity, charts should be drawn to reduce the need for narrative, some methods of flowcharting are described at the end of this chapter.

Section 7 – Suggestions and ideas. As the investigation proceeds it is likely that the system analyst will be thinking about possible solutions to problems that arise. Certainly he will start to come up against snags that will need to be overcome in the subsequent design phase. Notes about these matters should be written while they are still fresh in the mind; they should be filed in an orderly fashion so that they can be reviewed later. Eventually these notes can be worked into a statement of design objectives to be observed later.

Flowcharting Methods

There are a number of techniques of flowcharting and many individual preferences to be encountered among data processing people. It is to be hoped that some standards have been set out for the systems department by the systems manager. As far as this book is concerned, distinction is made here between flowcharts of clerical procedures and computer run flowcharts. These are the two main requirements to be distinguished in a systems investigation. For the most part these charts can be drawn using flowchart symbols specified by major international standards organizations.

Templates for drawing such symbols are usually available from computer manufacturers; details of these symbols are given in Fig. 2.8.

THE ASSIGNMENT BRIEF

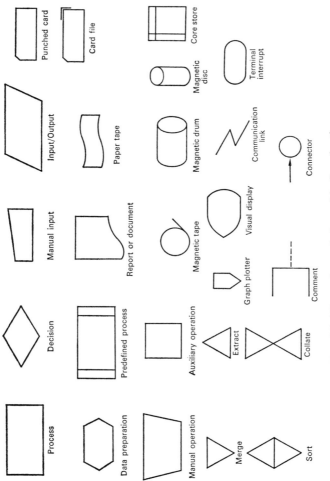

Fig. 2.8. The main symbols used in flowcharting.

SYSTEMS ANALYSIS IN BUSINESS

INVENTORY CONTROL SYSTEM FLOW CHART NO. 20

Order Dept. Logging and Data Control

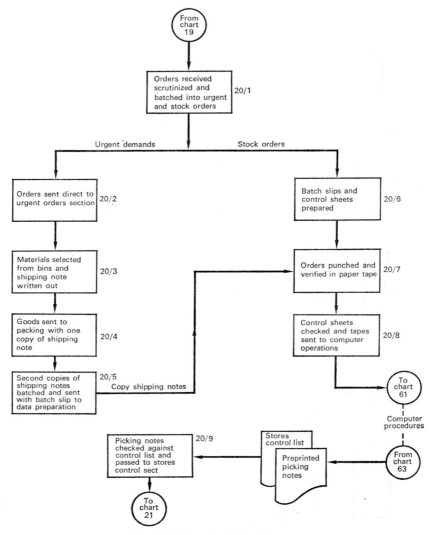

FIG. 2.9. Sample block diagram.

Block Diagrams for Procedure Charts

A block diagram provides a simple method for describing a procedure or system, it is a medium which reduces a problem to a number of definite steps.

THE ASSIGNMENT BRIEF

Procedures are relatively easy to follow when charted in this way; each major step is represented by a box which contains a brief description of the activity that it represents. Arrows connecting the boxes enable the flow of the procedure to be followed.

A section of a block diagram is given as a sample in Fig. 2.9. The top of each chart is headed with the project name, the date, name of the analyst, the particular procedure is identified, and a chart number indicates its relation to other charts used in representing the complete procedure. Cross-references to other charts are given at relevant points in the flowchart. In this particular example a cross-reference is made to a computer run-flowchart where data enter a computer procedure.

If necessary, such charts can be accompanied by narrative with cross-references to relevant box numbers on the chart.

Flowcharts of this nature should be drawn on standard sized paper, say A4; large sheets of paper are useful for wall charts but not so convenient for filing and handling, and should be avoided unless absolutely necessary. It is better to increase the number of pages needed for a procedure than to crowd too many details on to one page of a flowchart. It will be easier to use the chart if there is plenty of space on each page.

Computer Run-flowcharts

Computer run-flowcharts are used to represent computer procedures; each program is identified by a separate box, and symbols connected to each box indicate the number of files read or generated by the program and the nature of the peripheral devices used (Fig. 2.10).

A run-flowchart is usually produced for a suite of programs that one would normally expect to be run together at some specific point in the operation of a system, e.g. a daily invoicing run. Cross-references to other program suites can be given as shown in the example.

Such flowcharts are used in feasibility reports, system definitions, operating guides or manuals, etc. They represent a very simple method of showing the flow of information in a system, the relationship between programs, the files employed, their methods of updating, and reporting procedures. Cross-references can be made to clerical flowcharts where information enters or leaves the computer procedures.

ANALYSIS AND APPRAISAL OF AN EXISTING SYSTEM

Having completed the initial investigation and recording of details about the existing system, it is necessary to have a definite pause before embarking upon the design of the new system. This pause should permit an effective evaluation

SYSTEMS ANALYSIS IN BUSINESS

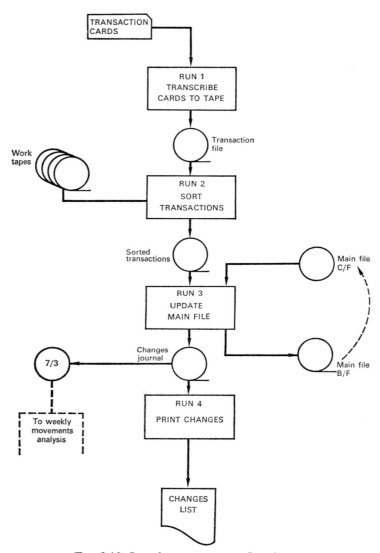

Fig. 2.10. Sample computer run-flowchart.

of the current system, which will probably give rise to comments and proposals under the following headings:

(a) The effectiveness of the current system in meeting its previously stated objectives, including some figures to show the level of performance attained;

48

THE ASSIGNMENT BRIEF

(b) The potential for incorporating other, as yet unspecified, benefits or functions within the system development plan;

(c) Opportunities that exist for making immediate improvements to the system by relatively simple changes to current procedures.

To make such an appraisal it is necessary to have in mind some specific objectives when starting the project, and, as we have mentioned earlier in this chapter, it is a mistake to embark upon a detailed investigation without definite aims in mind.

Measuring Performance

Let us refer once more to Fig. 2.4 where we can see some definite objectives stated; apart from the list of functions to be tackled, it is also necessary in this example to ensure compatibility with other nominated systems as well as achieve specific time requirements for handling throughput. The effectiveness of the current system in meeting these demands has to be established. In this case the present order processing system may be observed to take, say, ten days to deal with demands received from sources in the United Kingdom.

The analysis entails, therefore, a measurement of the average volume of demands to be handled in a given period and an appraisal of the loading that falls thereby upon the existing procedures. The costs of running the existing system have also to be established for different loads. Later on, any new designs will need to stand up to the same standards of measurement.

The appraisal should also attempt to evaluate the ability of the system to cope with future changes in working conditions or loads; for example, the analyst will have to consider whether the system can provide greater throughput perhaps by simply increasing resources at key points.

Effects Upon Adjacent Procedures

An understanding of the relationship between the current system and adjacent functions or procedures is also necessary. Perhaps the performance of the system being reviewed is constrained by the effectiveness of other systems. In the case of the order processing example, it may be found that too many items are not capable of being serviced because of stock shortages. The operation of the order processing system is therefore affected by having to back order these items, thus throwing additional loads upon some sections of the organization.

The current system may itself suffer from deficiencies which incur penalties in other system areas. All these facts have to be measured and evaluated by the analyst using his own powers of observation. In situations like these, it is often the case that different managers will place different constructions upon the relative performance of their departments, and the system analyst's independent appraisal has therefore to be supported by facts.

It is quite possible that the initial assignment brief did not include mention of certain functions that are closely related to the problem under review. The analyst may decide that the design of any new system should include consideration of these functions where there is a chance that the functions concerned will, if not included, lose the support of the procedures being reviewed. He should then check back with his manager to obtain approval before proceeding.

Identifying Simple Improvements

So far the analyst will not have considered the detailed design of any new system; his task to date is simply to evaluate the existing procedures. Indeed, it may turn out that the current system can, with some minor changes, be made to meet the expectations of management. This point is well worth emphasis here: too often one can see new systems developed, perhaps introducing expensive hardware and software resources, for achieving only marginal benefits. Some methods of assessing costs and performance are given in Chapter 3 when we discuss the feasibility of new systems.

Whether or not a new system is needed, there are probably a number of immediate advantages to be obtained by improving the existing procedures, and these should not be overlooked.

Reports Upon the Investigation

The objective of the investigation is to determine whether the existing system is capable of meeting its performance requirements both now and in the future. It is sometimes necessary to issue a report to the management outlining the findings in some detail, giving performance figures and explaining the logic used to arrive at specific conclusions. Certainly if there is any doubt about the performance required of the system, such matters should be identified and settled before conducting a feasibility study or detailed system design.

SUMMARY OF INVESTIGATION OBJECTIVES

1. The current system is reviewed to see whether its output meets the functional requirements of the departments concerned and whether its performance is up to the expectations of management and of other adjacent procedures that it supports.

2. If the current system is inefficient efforts must be made to express its performance in meaningful terms, and operational costs should be evaluated for different levels of loading that occur or that may arise in the future.

3. It may also be necessary to see if the system is able to cope with changes implied by new organizational policies or objectives, and by business or legal conditions affecting these policies.

4. Perhaps the system has to be reviewed in order to take advantage of new technological factors prevailing in the industry concerned.

5. The system may be required to generate more information or to achieve improvements in the quality or timing of information generated.

6. The good aspects of the current system must be identified and performance in this respect evaluated. Such characteristics should be noted for retention in any new system needed.

7. Perhaps the desired improvements can be attained by changes to the existing system: the investigation should indicate whether such changes are of a minor or major nature and for how long are such changes likely to meet future requirements.

Reviewing the Assignment Objectives

If the initial investigation calls for a more detailed feasibility study, or if a detailed system design is immediately indicated, the analyst must seek to reaffirm the overall objectives of the task. In the case of a feasibility study, the analyst should establish once more the bounds of the study, why it is being conducted, and what objectives are now being aimed at. This may call for a revision of the original assignment brief and its authorization once more.

If the next step is to design a new system it may also be necessary to prepare a statement of design objectives for inclusion as features of the new system. In any event the analyst must see that all subsequent design proposals are evaluated with the same intensity and vigour as is applied to the existing system.

Chapter 3

The Feasibility Study

WHY HAVE A FEASIBILITY STUDY?

In the previous chapter we have explained how an assignment brief is developed and how important it is to identify and specify the functions that are to be covered in a subsequent feasibility study. Having completed the initial investigation, the analyst will have gained a great deal of background knowledge of the problem area and may have already formed opinions about the future shape and structure of the new system.

If the problem is a relatively simple one, which can perhaps be solved by a straightforward batch processing application, it may not be necessary to start a full-scale feasibility study. This situation is more likely to occur in an organization in which there has been a tradition of data processing, and where there are several existing files of information which can be used to generate reports or maintain some other related procedures.

If, on the other hand, the organization is considering using a computer for the first time, or is introducing computer processing into a new problem area, it will almost certainly be necessary to conduct a more detailed study to evaluate the potential advantages to be gained. Computer systems often take many man-years to develop and have high development and operational costs to recover. Therefore a feasibility study should establish the best way of achieving the given objectives of the assignment and minimize the time and expenditure involved in implementing and running the new system.

Some problems are of a more technical nature or have certain characteristics which suggest that practical difficulties will have to be solved before embarking upon programming and implementation of a new scheme. For example, it could be that an organization which has used a batch processing computer for running their accounting and reporting procedures now wishes to tackle an application in which there is a functional need to supply data on-line to a number of departments located in different parts of the country. The nature of the problem suggests that direct access backing store, communica-

THE FEASIBILITY STUDY

tions terminals, and special communications software are needed. In such circumstances it is essential to conduct a feasibility study to ascertain the nature and capacity of the hardware and software resources needed, and the outline costs of developing, installing, and running the system.

The term *feasibility study* is perhaps a little misleading, since most business problems are capable of being mastered by a system of some kind. Modern man is continually concerned to rationalize the economic and social problems that surround him, and attempts to create an ordered environment within which predictions can be made. From the start it is evident that most business problems are amenable to the application of system logic. However, some solutions to problems are more practicable than others; some solutions are more easily implemented in a particular organization, some solutions are more costly to develop, some more costly to run. Perhaps time is the enemy in one case, and a solution has to be found that can be implemented quickly.

These constraints are part of the problem, and if the assignment brief has been correctly developed it may well include statements that will serve to identify them. There will naturally be a strong case for a feasibility study if an organization is breaking new ground in data processing work, but every computer system should be evaluated to establish the costs and benefits that will be incurred by its implementation. If an organization has spare computing capacity it may perhaps allow some applications to be developed without first determining these factors in detail, but in the long run a number of such applications may arise which, when considered in total, will occupy many of the organizations resources and yet yield little in increased efficiency or profitability.

A feasibility study should therefore entail the design of one or more solutions in outline and in sufficient depth to permit answers to the following questions to be stated:

(a) Can the objectives set by management be met by a new system?
(b) What advantages would a new system have over the present one?
(c) What are the costs of developing such a system?
(d) What are the costs of running a new system?
(e) How are the development and running costs related to the various functional objectives sought?
(f) Which of the objectives provide the best rewards *pro rata*?
(g) Can a program for phased implementation be established which will enable the more urgent objectives to be achieved quickly?

These are simple and practical considerations before undertaking any project whether it be in system development or not. An attempt is made to ascertain the opportunity costs of possible methods so that effective decisions governing future actions can be made.

HOW BENEFITS CAN BE EVALUATED

Increased Volumes or Throughput

The simplest benefits to measure are those that deal with the ability of a system to meet schedules and to handle increased volumes of data. Such benefits are easily demonstrated on paper and are usually attributable to increased power and efficiency of new hardware over existing equipment. It is sometimes the case that these benefits alone will justify the new system; but generally other related advantages will accrue, and they may not be so easily demonstrated. Methods for timing and evaluating computer processing are given in Chapters 6 and 7.

Reduction of Clerical Operations

A system may aim to reduce clerical costs by taking over many of the functions performed by people, and this in turn may reduce pressure on office space and equipment needed to support the clerical activity. These savings are often difficult to evaluate without doing a very detailed study of work in the department concerned. An increase in clerical productivity is usually possible only by deliberate efforts to transfer tasks to automated routines.

In practice these savings may prove difficult to attain because the clerical operative has to play an important quality control function in monitoring error reports, management reports, and documents generated by any computer system. In many cases clerical staff have to be developed and upgraded to meet the new demands placed upon them.

The collection, control, and preparation of input data have also to be evaluated before arriving at any conclusions about the quality and amount of clerical effort needed. To make such assessments a very detailed work study of the departments concerned is vital; if a reduction in clerical effort is stated as a main system objective, then the feasibility study should be conducted with this requirement firmly in mind.

A feasibility study should evaluate the numbers and grades of staff needed in these supporting roles, and must present figures for comparison with the establishment currently needed for the existing system. If the proposed system has other advantages, it may well be possible to show that the new system uses staff in a more creative and productive manner.

Improved Internal Communication

It is common to find organizational structures in which separate departments duplicate functions or produce conflicting information. Usually these conditions arise because local systems have to be set up to serve different but similar needs. Departments often establish files and procedures which meet their own local problems, and the quality of the data is relevant only to their particular operations. A computer system provides a great opportunity to eliminate or reduce the local data processing operations and to provide a common data base for the separate units in the organization. It may be possible at the same time to effect organizational changes and improve communications within the group.

Improvements in the Quality of Information

Under this heading come many benefits which can generally be classified as improving the control of the whole business cooperation. Claims of this nature are easy to make but more difficult to justify. Improvements may be in the accuracy and the timeliness of information thus permitting more timely decisions by management. Wherever possible these benefits have to be assessed in relation to particular functions performed within the management group, showing how decision-making is aided in particular cases by outputs from the new system. A system that simply collects and collates relevant facts has no intrinsic value of its own, and must be related to the functions performed by certain managers or departments.

Reduced Interest Charges

Any system which improves the control of operations may serve to minimize stocks of finished or unfinished goods and raw materials. The interest charges associated with borrowed capital to finance production and stockholding may therefore be reduced. Improvements in the efficiency and accuracy of invoicing routines may similarly improve cash flow. These are all valid and important benefits to be derived from a new system, but they should be sought by a careful evaluation of the potential of the new methods. The feasibility study should attempt to draw sensible and practical conclusions by pilot schemes or analyses developed manually simulating the logic and selective power of the proposed computer routines. If assessments of this nature are not made carefully and with due regard to actual conditions prevailing in the departments being studied, some surprising results can materialize when the system eventually operates in the live environment.

A case in point relates to an organization which introduced a computer system to control stocks on the understanding that one advantage would be to reduce stock levels and thus minimize associated interest charges and the warehouse space occupied. In fact, the eventual result was to increase stock levels because, using the old system, items were underprovisioned; to offset this there was, using the new system, an increase in service to the customer which encouraged more business, and a reduction of shortages which improved general operating efficiency. Also a more rational policy for purchasing was established which enabled better discounts to be achieved from suppliers, and the administrative costs of placing purchase orders were reduced.

It is difficult and pointless to make sweeping predictions of the likely results of a new system; too much optimism does not help the analyst's credibility in the long term. But current conditions and recent history can be studied so that statements can be made to demonstrate the advantages that will be obtained if particular methods of operation are secured.

Improved Relations with Outside Groups

A new system may improve relations with customers and/or suppliers through improvements in the speed and accuracy with which transactions can be processed. These improvements may enhance the organization's competitiveness or, in the case of some government agency, improve service to the public.

Summary of Benefits

It has only been possible to indicate above some of the ways in which benefits can be obtained. There are many other advantages afforded by systems, and it is for the analyst to examine the objectives and likely effects of a new proposal in his particular organization. As a general rule, these benefits can be expressed in terms of an increase in net profitability or efficiency. To appreciate the importance of these claims there must also be a schedule of the costs incurred by running or introducing the new system. In this way the management will be able to assess the value of continuing with the detailed design of the system, bearing in mind other projects that are under review and the resources currently available for the work.

ASSESSING THE COSTS OF A NEW SYSTEM

The costs of any new system may be considered under two main headings:

(a) Running costs;
(b) Development and implementation costs.

Running Costs

Running costs are more easily determined from the systems outline: development costs, although basically of a similar nature, are governed by current conditions, and we shall say more about them later. Both of these present difficulties, particularly in a large and complex system involving many departments and functions. For the most part, it is possible to break costs down into further categories under the following headings:

(a) Staff;
(b) Equipment;

THE FEASIBILITY STUDY

(c) Stationery and supplies;
(d) Physical space and related overheads.

An analysis of the proposed procedures under these main headings will identify the costs associated with particular functions or procedures.

Staff Costs

The staff costs can once more be broken down for further analysis as follows:

(a) Clerical operations and data control;
(b) Data preparation;
(c) Data processing support.

The new system can be defined as consisting of a number of procedures which require specific operations to be performed. Each operation has to be studied to determine the processes that must be completed, and for each process an average time has to be determined. This is not always easy because some clerical operations require judgement and discretion on the part of individuals performing them. Nevertheless, a careful study of these operations in consultation with the relevant members of the organization will develop a better understanding. Problems may also arise because individuals have to perform several functions which may overlap, or may be dependent upon some other operations that fluctuate in volume or intensity. In these cases the work has to be measured thoroughly by a methods analyst.

Clerical Operations and Data Control

Clerical functions such as filing, recording, and typing are gradually being absorbed by computer systems, in which many of the intermediate clerical operations become redundant. However, this will not necessarily minimize the number of staff required but will entail a reorganization of clerical resources to support the new system. Difficulties often arise in installing a new system because the staff concerned have not been organized to run an old and a new system in parallel. This stage can be conducted satisfactorily only if the new tasks have been evaluated and accepted by the departments concerned and if establishments have been arranged to cope with the work loads arising. In supporting computer systems, clerical activities are usually centred around:

(a) Input collection and control;
(b) Maintenance of control accounts;
(c) Dealing with error reports;
(d) Inspecting quality of output;
(e) Distribution of output.

These functions may be performed by the data processing department or a user department for whom the system has been designed. Usually these functions are shared between the departments concerned.

Input collection and control may entail the receipt, scrutiny, logging, and batching of transactions. The people who do this function may also be responsible for maintenance of control accounts to verify the correct operation of the system and for inspection of output to verify accuracy or quality of production.

Data Preparation

This entails the receipt of data, perhaps on original documents, and its conversion into some input media acceptable to the computer. In the case of punched cards or paper tape this involves punching and verifying. We shall return to this subject in Chapter 5 with more detailed analysis of the operations involved.

Data Processing Support

As far as possible, a computer system may be designed to run with as little human intervention as possible. Computer operations should entail a predetermined set of activities with all decisions based upon logic specified in the system. Where this is so, the operating activity can be determined from the equipment loading for the new system. However, it is not unusual to expect some work to be accomplished in preparing the job and in co-ordinating the computer operations according to the user's wishes at particular times. Work in this category may include:

(a) Control of computer files and data security;
(b) Preparation of jobs for running including the specification of run time conditions and parameters;
(c) Program maintenance.

Equipment Costs

A computer system entails the use of computer hardware and ancillary equipment used in data collection and transmission. Most computers are nowadays multiprogramming systems, and it is unlikely that one application will absorb all the equipment resources at one time. It therefore is usual to estimate costs by allocating charges to particular jobs according to the number and type of units used and to the duration for which a particular job occupies the units concerned, the costs for various units being standards set by the operations manager.

Some types of equipment are acquired specifically for a particular task; for example, a visual display terminal may be provided to cover one particular function. Sometimes other general office machines and facilities are needed

THE FEASIBILITY STUDY

for supporting a system, e.g. teletype facilities or storage cabinets. If the new system is to be costed for comparison with the current system, such items must be considered.

Physical Space and Related Overheads

The costs of physical space and such overhead factors as heating, lighting, and general office services must also be considered, but there is a danger of wasting much time and effort on each particular study, and this should be done in a standard manner. If the functions being tackled are of a fundamental nature and imply the establishment of new buildings or special environment for a new computer, then clearly all costs should be analysed in detail before approval of the project is granted; the costs can then be allocated to departments according to their utilization of the resources concerned.

The analyst would not usually be expected to cost individual projects in such detail, in a feasibility study, but he should do it in such a manner that it bears comparison with the evaluation made of the existing system. It is usually satisfactory to identify the main resources needed for a system, and one would expect that the data processing management will have provided standard cost factors that can be used to effect comparisons.

Costs of Development and Implementation

Development costs are not continuing, but they may be very significant, and development may utilize resources that are in short supply. For the most part development will entail the use of the same type of resources described above, but will include also the functions of:

(a) Detailed system design;
(b) The writing and testing of programs;
(c) Documentation for the system.

These activities must be broken down and an outline schedule produced so that it is possible to give in the feasibility report an estimation of the lead time and staff needed for getting the project operational.

These topics are covered in later chapters of this book and in Chapter 9 a proposed method for controlling and scheduling the whole systems development function is given.

THE DILEMMA OF A FEASIBILITY STUDY

Without a feasibility study an organization may fail to appreciate the difficulties or the potential of a systems project until all the design work has been completed. The feasibility report should provide an adequate means of obtaining a reasoned view at an early stage. On the other hand, if the design is not completed in sufficient detail it may be difficult to appraise the relative costs and benefits of a new system, so the ground has to be covered thoroughly.

In these matters guidance is necessary from management at the assignment stage to identify the main factors that require attention, and the data processing management must play an active role in setting the objectives and anticipating problems.

An experienced analyst will be able to use his own judgement in evaluating the importance of matters as he goes along. If he thinks the assignment is wrongly based in that it overlooks important issues, he must ensure that these matters are brought to light in the feasibility report or during the initial investigation.

The overall objectives are to estimate the benefits and costs and to identify major resources needed so that early efforts can be made to marshal them when the project is developed further. Major problems can be isolated, and perhaps further studies can be indicated where it is necessary to evaluate the efficacy of particular techniques or equipment.

A proposed structure and format for a feasibility report are given in Chapter 10, where the question of documentation is considered.

Chapter 4

The Approach to Systems Design

FIRST REVIEW THE OBJECTIVES

The design phase is perhaps the most satisfying aspect of the analyst's task. To some extent he has to disengage from the current restrictions that the organization and its procedures impose and allow his mind to range over the functions that have to be performed without prejudice. In the end, however, his solution has to be capable of implementation within the structure of the organization, and practical problems and human considerations are inevitably part of the overall task. Many systems projects cannot be said to be completed at any particular moment in time, since one phase of development often follows close upon another. What is important is that each step is made in the right direction – when an analyst undertakes any project involving the primary systems within a company he is changing the character of the organization, changing the tools and means by which the primary aims of the group are achieved.

The design phase of any new system or subsystem rarely has a definite beginning. It perhaps begins to take shape in the minds of line management who are seeking methods to improve operations on a day-to-day basis and are consciously setting aims and performance levels needed by a new system. These aims get transferred to the analyst when he first receives or develops the assignment brief, and throughout the initial investigation and the feasibility study he may start to commit to paper potential features of a new system.

When the feasibility stage has been completed, the analyst must review these previous ideas to see if they still have the same significance. Furthermore, he should seek to reaffirm the objectives of the systems project and review the outputs required from the system in content, format, and frequency of production.

When the design phase begins in earnest, both the analyst and the line management will have learnt much more about the operations that they are trying to control, and will be better placed to finally specify the output details

61

required by the system. In its simplest form the design stage entails working back from these output specifications through the following steps:

(a) Determine the elements of data required in output, their structure, degree of accuracy and frequency of production;
(b) Develop a file structure in which these elements can be stored or generated to produce the required output;
(c) Determine which data elements are to be collected to provide the information, determine their source, format, structure, and frequency of collection;
(d) Develop a file processing structure which enables the data to be processed to meet the specified output objectives, with efficiency and at a relevant cost.

Relating Output to Functional Needs

The steps given above will be useless if the output specifications are not conceived properly in the first instance. Any analyst should be forgiven if he appears to labour over this aspect, for it is the keynote to the eventual success of all his efforts. A full analysis of the functional use of the output must be undertaken; for example, if a report is to be used to make some specific decisions on a day-to-day basis the following points should be considered:

(a) Identify the decisions that are to be made;
(b) Identify the data elements that must be presented to make the decisions;
(c) Consider the timing of decisions and their relationship to the operations being controlled;
(d) Determine the response required of the system in order that the conditions of (c) above can be fulfilled;
(e) Determine the accuracy needed in the generation of results;
(f) Assess the volume of decisions to be made and the number of people who must make them;
(g) Establish a system of reporting so that only exception conditions requiring action are presented;
(h) Determine the medium and format of output that will best serve the needs above.
(i) Specify day-to-day checks to be performed to affirm the accuracy and quality of output relevant to the functions being controlled.

The best way of verifying these details, once they have been initially drafted, is to spend some time in the user department actually doing the various tasks that utilize the output. Many detailed aspects of the job will be revealed in this way, and small improvements in the specification can be made which will significantly improve the efficiency of the output. There is certainly a danger

THE APPROACH TO SYSTEMS DESIGN

in producing results which have been specified by one man, e.g. the head of the department or a spokesman for the staff concerned. The analyst must be quite firm about meeting the people who are doing the job concerned so that he can observe and evaluate the different view points.

SPECIFICATION OF OUTPUT

Output may be in a variety of media: e.g. a printed report used by management, a punched paper tape to control a machine tool, a display on a visual display unit, an intermediate file for input to another system, a functional document such as a purchase order, etc. It is vital to agree formats in detail at the commencement of the design phase because minor changes, if not anticipated, can have far-reaching effects on programs and file structures.

Output formats should be drawn up on layout charts and should be agreed with the users who must formally signify their approval before detailed file and procedure specifications are developed. If possible, sample layouts should be produced in the appropriate media and format – printed samples generated by dummy print programs on a line printer are much more effective than handwritten or typed samples.

The individual data elements required in output must be examined carefully to ascertain their format and magnitudes. Totals generated by the system are particularly important, allowance must be made for the maximum possible field sizes that can arise. The analyst should perform research upon existing files and documents, and consider peak volumes for the new system by looking to future growth and expansion of the functions being controlled. He should not rely on the word of individuals in the line departments in these matters; they may not be accustomed to dealing with the various data elements in this way.

Future Expansion of Output

We have discussed how output must be viewed in relation to the organizations needs; by the same principle it is necessary to consider the future aims and objectives of the group concerned and attempt to build flexibility into the reporting procedures to meet future needs.

Output formats and particularly print files and programs, should be conceived to permit future changes in the requirements with as little delay as possible. A little foresight in dealing with certain elements of data is needed – a company may be expected to increase the number of products that it markets, to change its organization, and thus related cost centres in its accounting procedures, or to expand the channels of distribution and so on.

SYSTEMS ANALYSIS IN BUSINESS

Approval of Output Formats

The output may well be the only aspect of the system that the user sees. He will probably not fully understand the procedures that support his output or understand the relevance of adjacent systems and clerical procedures. In verifying the output and signifying his assent to the format, content, and frequency, he is in effect authorizing the system. The analyst has, therefore, to ensure that the output is fully understood by those who are going to use it, and the general aims of the system should be agreed both with line management and with executives at the policy setting level.

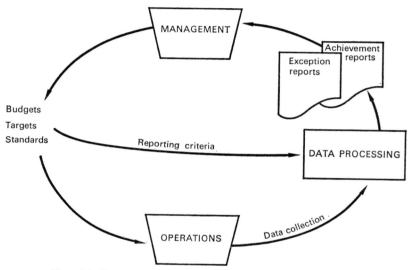

FIG. 4.1. Data processing in management control systems.

Approval generally takes place initially at the feasibility stage, but in some cases a lengthy feasibility study will be unnecessary. Before implementing any procedures or completing detailed program specifications, the analyst should reaffirm the overall system objectives by presentations to users and their management in which the relevance of output is discussed openly and in detail. It is important to ensure that the relevance of output to current and future objectives is appraised.

Formats should be approved formally by signature of the users and the systems staff concerned.

Exception Reporting

Some computer systems are installed specifically to produce large numbers of output documents, e.g. insurance policy renewal statements, pay slips, and so

THE APPROACH TO SYSTEMS DESIGN

on. In such cases each document fulfils some well-defined functional requirement. However, management information systems should not generate large output volumes, and those that do must be suspected of failing to produce relevant and meaningful results.

Computer information systems used to improve control over operations, must be regarded as being part of a man–machine system in which action is taken within a control framework, as shown in Fig. 4.1. The system collects information about an operation, collates and assembles data to present performance statistics, and the management observing these statistics decides whether operations meet required standards. Changes to standards or to the resources and methods used in the operations are then made and results observed once more.

This must be a continuing process and one in which the system does not require to list all the events and transactions that arise, but simply reports relevant details to the managers responsible for various functions within the overall operation.

For example, in a manufacturing organization, the divisional manager may wish to know, once a week, whether his producing units are working to the agreed production schedule, and how their expenditure is related to budgeted expenditure on production during the year to date. He will want a brief summarized report each week from the system which will highlight persistent trends at each manufacturing centre and perhaps highlight products which are consistently presenting problems.

The line manager responsible for a particular production program will probably need much more detailed analysis of particular products, perhaps illustrating the labour and material costs against budgeted costs in the year to date. Perhaps an abnormal number of a particular product are failing at quality testing; the line manager will want to set standards for this function, be informed when they are violated, and be presented with an analysis by product by reason for failure. He will not want to look at those items which are being produced within defined limits, and will probably not be very concerned to examine details of products which exhibit an abnormal condition in one week only. He will, however, want to see details of items consistently outside nominated bounds, and will probably wish to change reporting criteria as new products or conditions arise.

The techniques for extracting and presenting relevant facts in such systems must be based upon flexible principles; perhaps parameterized routines should be developed to facilitate changes at run time when new criteria for reporting are established.

Output and Control of System Errors

The analyst must anticipate errors and illogical situations that can arise in the operation of his system, and should design computer routines to identify these conditions as and when they arise. All error reports have to be dealt with by users and the following principles should be remembered when designing error prints:

(a) Attempt to foresee sources of error and develop procedures to isolate them;
(b) Design error reports with care, so that they provide suitable diagnostic comments and the relevant facts to facilitate amendment;
(c) Agree the format and content of error lists with those who are going to use them;
(d) Only report conditions on which action is required;
(e) Set a tolerance level for volumes of error outputs and review systems if these levels are consistently exceeded.

DATA COLLECTION AND FILE PROCESSING

Once the output and aims of the system have been finally established, the analyst must design the computer procedures and the data collection and preparation activities needed to support the system. These are specialist tasks in which the systems analyst is able to display his own creative talents. In this book, Chapters 5, 6, and 7 are devoted to explaining in some detail the main principles and techniques involved; in the present chapter we shall simply try to identify broad guidelines for these aspects of the work.

Analysing Data Collection Needs

The data elements required to produce a given output can be considered in three major categories:

(a) Derived data which are stored or generated by the system;
(b) Raw data which are collected from external sources and which may appear directly in output or be used to provide derived data;
(c) Descriptive data.

In a stock control system, examples of derived data are, *stock value, stock quantity*, and *written-down value*; whereas raw data examples include *quantity ordered* and *quantity delivered*. Descriptive data includes data elements such as *price, name, address, account code*, and *item description*.

Thus the analyst can identify the data which must be collected and stored in files or be generated from file information for output. The frequency of the output required also determines the frequency with which data is collected, files updated, and output reports generated.

For example, to produce a weekly payroll for hourly paid staff it may be

THE APPROACH TO SYSTEMS DESIGN

necessary to collect details about staff attendance during each week, the number of standard hours each has worked, the number of overtime hours worked, the type of jobs performed throughout this time. These are items of raw data needed for the payroll, and it is likely that these will be collected hourly or daily against each staff number in the day-to-day administration of the department concerned as individual jobs are completed. The data may need further preparation before entry to the computer procedures, but it must be ready for input to the system at least once per week.

Within the file system this raw data is processed and matched with relevant descriptive information needed to generate the payroll. The rate for each job must be stored, each individual's name, his tax code, other fixed deductions from salary, and so on. The point to be observed at this stage is that descriptive data is rarely of a permanent nature; staff leave and join, rates change, fixed deductions vary, and tax codes can be altered by an individual's circumstances. Therefore routines must be established to maintain the accuracy of these data elements in the files, and these routines must also be geared to the system's output frequency.

The collection of raw data and additions or amendments to descriptive data can be related to particular functions and procedures. These procedures must be developed to provide a ready source for this data in accordance with the accuracy and response required for the system. Perhaps documents and procedures already exist to cover the requirement, and these will have been brought to light in the initial investigation or feasibility study.

Determining Structure and Characteristics of Data

The data to be handled by a system has to be studied, and a detailed specification of the individual data elements completed. The design of input and output formats and the file and record layouts can be finalized only when the size and characteristics of the data elements have been settled. The maximum size of individual items collected, generated, and output by the system has to be ascertained; e.g. the number of characters in input and output formats, the number of words or bits occupied in internal memory. The range of values that particular elements may take can also be important, particularly where items of data are examined logically to determine courses of action in processing and handling.

Fields to be used when performing calculations have to be studied so that allowance can be made for the range of values that the answers may take: the degree of precision obtained in performing calculations may also be affected.

The detailed structure of existing or new codes has to be determined at an early stage, and the potential for expanding the code ranges to cater for future needs must be considered. During input processing it is usual to examine fields to see that they fall within specified sizes and formats allowed by the system. Other characteristics helpful in carrying out validation should be identified. Fields used as keys in sorting and arranging operations have to be studied to ensure that a desired sequence can be obtained.

The following list is a useful guide when considering the degree of detail necessary for specifying data elements:

   (a) Data name – description of element;
   (b) Source – input record type;
   (c) External format – numeric, alphabetic, alphanumeric;
   (d) Size – number of characters;
   (e) Range – range of values, e.g. greater than 270 less than 480;
   (f) Internal storage format – character or binary, right or left justified;
   (g) Code reference – refers to separate code book or list.

The characteristics of existing items will have been ascertained during the investigation and analysis of existing procedures and documentation. This information must be continually reviewed as the analysis and design proceeds, and at all times be examined to ascertain the potential for change bearing in mind the environment of the system and the organization.

Some Aspects of Data Collection

Having identified the data elements that need to be collected and related them to particular functions and procedures, the analyst has to determine the volume and frequency of these transactions and specify the methods for collecting the raw data. As first objectives, the analyst should aim to meet the following requirements:

   (a) Capture data as closely as possible to the time and place at which the events occur;
   (b) Attempt to design data collection methods such that the events and functions concerned are dependent upon correctly recording the data elements;
   (c) Avoid or minimize any unnecessary transcription of data from one medium to another;
   (d) Use data collection techniques that minimize potential human errors, and employ automated methods wherever practicable;
   (e) Make certain that the work concerned with data collection and preparation is correctly measured, evaluated, and accepted by the user departments concerned;
   (f) Establish control procedures which will serve to verify the accuracy and completeness of data entering a system.

THE APPROACH TO SYSTEMS DESIGN

FILE DESIGN

File design entails, principally, deciding upon the contents and structure of individual records and their arrangement and grouping into separate files, and the relationship between files. The nature of the application itself has an overall influence on these matters, and determines the type of file storage media needed for a particular application. Generally speaking, large files are accommodated either on magnetic tape or on a direct access storage medium such as magnetic discs or magnetic cards.

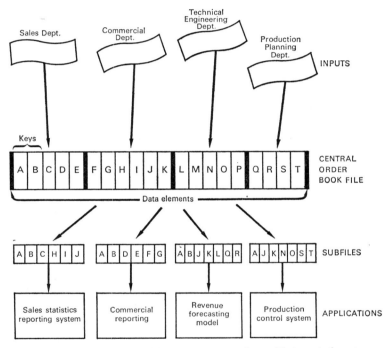

FIG. 4.2. Use of a central file to serve a number of integrated systems.

These media have distinct technical characteristics which suit particular application needs, e.g. magnetic tape is a serial medium and is suitable only for batch processing; direct access devices provide a more expensive form of storage, and perhaps more complex processing problems, but can also yield a faster response. The technical merits of these basic media are dealt with at length in Chapters 6 and 7.

The overall attitude within an organization to its data processing problems will also be reflected in the file design. If the group concerned is developing applications separately to tackle particular problem areas, then it is unlikely that the files needed for one application will be unduly influenced by those needed for another. If, on the other hand, systems are being developed as a set of integrated procedures in which several applications share common information, then files will be interdependent. Fig. 4.2 represents a likely approach in the latter case: here it is assumed that one major file is used to satisfy a number of applications – subfiles being created from it as and when needed to fulfil particular application needs.

SYSTEM CONTROLS

When a company commits its vital records and procedures to a computerized file processing system, it will want to be satisfied that the output generated by the system is accurate and reliable. It is possible, if adequate precautions are not taken, to cause errors in data processing on a large scale from relatively trivial events. Even the most reliable and thoroughly tested computer routines can produce errors if they are provided with incorrect or incomplete data to process.

Precautions are taken generally at the following levels:

(a) By the computer manufacturer who will include self-checking systems in the operation of the computer hardware and software;
(b) By the systems analyst who must build a system of control accounts into the computer procedures for each particular application;
(c) By the users of the system who must also maintain compatible control accounts for verifying the correct collection and preparation of input data;
(d) By the internal (or external) auditors who must agree, initially, the methods adopted for tackling a particular problem, and who must subsequently check for correct operation of systems and procedures.

In this chapter we will not discuss the hardware and software controls, since these are dealt with in Chapter 6 under the general heading Security of Data. Those who are particularly interested in hardware facilities should refer to a relevant publication dealing in detail with hardware concepts.

File Processing Control Accounts

A file processing system should maintain a system of control accounts to verify that all records have been processed, and an overall check upon the accuracy of processing operations. The figures for entry into these accounts should be generated from run to run, so that a trail can be followed to vet the correct operation in each file processing cycle.

Assume, for example, that a very simple system has three runs, handling

THE APPROACH TO SYSTEMS DESIGN

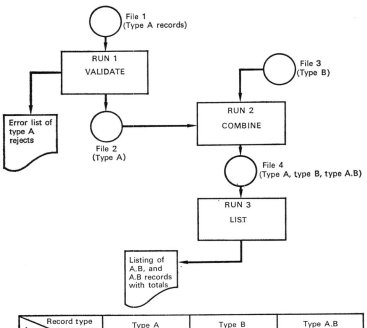

Record type → ↓ Run No.	Type A		Type B		Type A.B		
	Count of records	£ value	Count of records	£ value	Count of records	£ value	
1 Input	500	21,915	—	—	—	—	
Rejected	25	2,240	—	—	—	—	
Carried forward	475	19,675	—	—	—	—	Derived totals
2 Brought forward	475	19,675	10,302	175,899	—	—	195,574
Combined	400	15,075	400	6,658	—	—	21,733
Carried forward	75	4,600	9,902	169,241	400	21,733	
3 Brought forward	75	4,600	9,902	169,241	400	21,733	Computed totals
Output	75	4,600	9,902	169,241	400	21,733	195,574

FIG. 4.3. Simple control accounts.

two basic record types as shown in Fig. 4.3. In run 1, type A records are edited and validated, the invalid records being rejected and listed for inclusion later. In run 2, type A records are merged with a file of type B records, and where keys are matched the two records are compressed into output records known as type A.B. Finally, in run 3, the records are listed with final totals.

A system of control accounts as shown in Fig. 4.3 will serve to check the accuracy of data processing activity in the three runs – a tally of the number of records handled by the system is kept, and a value field appearing in all record types is summed to give a control total for checking through each stage.

A simple procedure such as this provides a reassurance, even when quite complex processing activity takes place in the suite of runs concerned. The value in maintaining such accounts is, however, nullified if the computer operations or data control staff do not check them. It is, therefore, helpful to request that totals generated by the programs are written down on a single sheet of paper designed for the purpose, or if the accounts are to be generated completely on the computer, ensure that the quality control staff sign and file these documents in an orderly fashion. It is useful to enter manually derived totals for balancing directly on to the control accounts.

Data Collection Control Accounts

There is a need to prepare and control data entering a system to check the completeness of input. Where an automatic data collection technique is used, then record counts and totals should be generated as transactions are recorded. For conventional punched-card or paper tape preparation procedures, the following steps should be observed:

(a) Collect and batch documents and assign batch numbers;
(b) Derive a count of entries or documents and, using an adding machine, develop control totals for each batch;
(c) Pass batches to the data preparation centre who should count batches and check batch numbers;
(d) Punch the data into input media (e.g. cards);
(e) Verify punching and correct errors;
(f) Using a tabulating device, read media and check number of records per batch and overall control totals;
(g) Input to computer processing and list all batch numbers and control totals per batch, indicating records accepted and rejected;
(h) Carry forward input accounts to file processing accounts.

Control accounts should be as brief and simple as possible to maintain; and all users and computer operations staff should be instructed in the use of the accounts and their importance to the operations dependent upon the system.

Control Systems and Auditors

All computer systems must satisfy the auditors of the organization concerned and it is essential to consult the audit authority at an early stage to be certain that adequate thought is given to the routines needed. If possible, a member

THE APPROACH TO SYSTEMS DESIGN

of the audit department should be included in design project teams, particularly where company assets such as raw materials and finished stocks are concerned.

The audit authority may include the following functions in their terms of reference:

(a) To vet that a system is able to achieve its stated objectives;
(b) That it conforms with accepted financial, legal, and business practice;
(c) That it uses resources efficiently;
(d) That the company's interests and assets are protected.

These functions can pervade the whole area of activity within the organization, and the extent to which the audit department becomes involved may well differ from one organization to another. Nevertheless, it is important to get all documentation and procedures vetted by them before programming and implementation begins.

Once a system is put into operation, the auditors will need to check that the system works correctly on a day-to-day basis. Not only will they wish to check clerical procedures but will need, also, to conduct program tests on live systems. The following activities may be necessary in this process:

(a) The audit department will need to approve all programs and subsequent amendments to programs;
(b) Maintain test packs of data to be used on-demand to check the results produced by live programs;
(c) To maintain copies of programs which can be used for comparison with those in use, perhaps having their own general purpose software developed for the purpose;
(d) To reserve the right to examine the console log to check operating instructions given to programs at run time;
(e) Select random samples of data from live files on demand and to examine history files showing previous actions.

Contact with the auditors is essential at every stage if intelligent arrangements for carrying out these functions are to be made, and systems analysts should maintain close liaison with the auditors as part of their day-to-day work.

Chapter 5

Data Collection and Preparation

THE IMPORTANCE OF DATA COLLECTION

One of the potentially weakest links of any computer system lies in data collection; particularly is this true of business applications in which large volumes of transactions have to be collected from the day-to-day operations of an organization. The expression 'garbage in, garbage out' is well known in the data processing business. It means simply that if you put bad data into a computer system the results will also be bad and probably unusable.

A computer itself is a very expensive tool. Furthermore, major computer applications often entail just as much expenditure on the software and supporting services as upon the original hardware. This expenditure is wasted if the methods and procedures used to collect data are inefficient. Computer projects can founder on relatively trivial problems, and poor control of input data is a common ailment.

The analyst's job is to design procedures that will ensure that data can be collected and prepared accurately, efficiently, and in good time for the system. The procedures have to be understood and be accepted by the people who will eventually have to run them, and the volume and nature of the work must be studied and evaluated to make sure that it is within the capacity of the departments concerned.

The traditional methods of data collection entail copying details of transactions from original source documents (e.g. advice notes, dispatch lists) on to some computer input media such as cards or paper tape. This procedure might entail some or all of the following steps:

(a) Collection of source documents and arrangement into batches;
(b) Generation of control totals from source documents and entry in manually prepared control accounts;
(c) Punching of source documents into input media;
(d) Verification of input media against original source documents;

DATA COLLECTION AND PREPARATION

(e) Checking control totals to see that all batches have been punched correctly;
(f) Checking of batch totals and entry into control accounts once more;
(g) Input to the computer procedures and the generation of control figures in the input program for checking against manually prepared control accounts, thus checking that all transactions and batches are accounted for.

This whole process is very time-consuming and may itself give rise to errors if stringent checking of control figures is not carried out. If the original source documents are not in suitable format for punching and verification to take place efficiently, then code clerks might be employed to transcribe data from source documents on to punching sheets as soon as the source documents have been collected and batched. This is yet another time-consuming and error-prone task, and the analyst must try to avoid it wherever possible.

The object of many computer systems in business organizations is to minimize overheads and to reduce the drudgery and boredom associated with routine clerical activities.

The traditional data preparation methods are themselves sources of boredom – much of the work is repetitive and unappealing to individuals. One has only to consider the time and resources occupied in undertaking this work and to see rows and rows of paper tape or punched-card preparation machines alongside a computer suite to realize how anachronistic these methods are.

Every effort should be expended to cut out intermediate operations, particularly those which are based upon a non-creative use of labour. Techniques of data collection are improving all the time, but there is no doubt that it is the one area of data processing technology which will continue to defy rapid progress. A systems analyst is advised to look long and hard into this activity: he must use all his ingenuity and knowledge of equipment and methods available to capture input data.

This weak link between man and the computer will always be the most vulnerable point of any computer control system; a systems analyst has to carefully appraise the effectiveness of new techniques yet not be deceived by their sophistication. In research institutes today, devices are being developed which can recognize a range of sounds made by humans and interpret them in an intelligible manner. Whether or not such machines become commonplace remains to be seen, but it is unlikely that they will greatly simplify the problem of man–machine communication: probably there will be greater potential ambiguity.

In this chapter some of the more well-known methods of data collection are described – each has a part to play in this field and can be considered as a candidate for the satisfaction of any particular problem. It is important to remember that data collection or preparation has no value in its own right – it is a means to an end. The aim is to convert data from a form which humans can recognize and interpret, to a form which the computer can handle. Simplicity is important, accuracy is essential, but cost and speed are also vital considerations.

PUNCHED CARDS

Punched-card equipment has been used for data processing since about 1890. Nowadays, except in very small organizations, punched-card machines are being replaced by electronic computer systems or are being used in a subsidiary role to computers. The great era of punched-card sorters, tabulators, and calculators began to close from about 1960, but punched cards are still frequently used as a means of input to computers.

The standard 80-column punched card is about $7\frac{3}{8}$ inches by $3\frac{1}{4}$ inches and about 0·007 inch thick. The card is divided into eighty columns numbered from 1 to 80, and each column contains twelve possible punching positions (Fig. 5.1). The twelve positions include nine numeric positions which are labelled 1, 2, 3, 4, 5, 6, 7, 8, and 9: a single hole may be punched in any one of these positions to represent a single numeric digit. The other three positions in a column are usually known as 0, 11, and 10: these are referred to as upper zone positions; any one of these can be used in conjunction with a numeric digit to represent an alphabetic character.

Up to 48 separate characters can be recorded using zone and numeric punching mentioned above. Some card codes permit three holes to be recorded in a single column and a range of 64 characters become available including numeric digits 0–9, alphabetic characters A–Z, and a wide range of special symbols (Fig. 5.2).

Fields of data are represented by a number of adjacent card columns; thus a field may be from 1 to 80 characters in length. For a particular job, fields are normally fixed in size, but they can be of variable length in which case special end-of-field markers may have to be punched to denote the bounds of each field.

Other sizes of card have been used in the past, and a systems analyst may come across some mechanical systems which employ 21-, 45-, 65-, or 90-column cards. Generally speaking computer manufacturers do not cater for these systems as a computer input media, but may be prepared to develop special input or output devices for users who have become very dependent upon such cards.

Punching of Cards

Punched cards are prepared directly from source documents: an operator

DATA COLLECTION AND PREPARATION

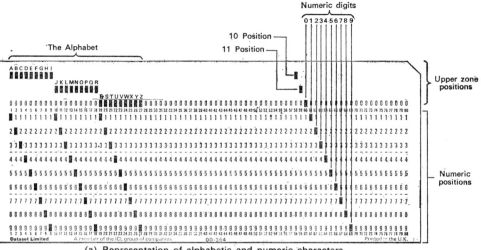

(a) Representation of alphabetic and numeric characters.

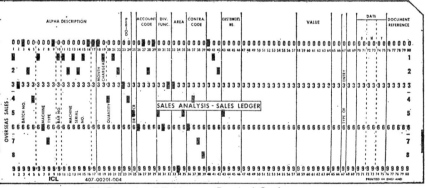

(b) Example of Fields on a Punched Card.

FIG. 5.1. Samples of 80-column punched card.

reads the documents and uses a keyboard to enter details to a machine which punches each card column by column. Punch machines may be simple mechanical devices known as *hand punches*, in which cards are loaded and ejected manually. Automatic punches are more commonly used where large volumes of data have to be prepared. They consist of a card hopper, into which several hundred blank cards may be initially placed; and a card rack,

SYSTEMS ANALYSIS IN BUSINESS

64-Character code

Character	Punching	Character	Punching	Character	Punching	Character	Punching
0	0	E	10,5	U	0,4	:	10,5,8
1	1	F	10,6	V	0,5	(apostrophe)	10,6,8
2	2	G	10,7	W	0,6	!	10,7,8
3	3	H	10,8	X	0,7	[	11,2,8
4	4	I	10,9	Y	0,8	$	11,3,8
5	5	J	11,1	Z	0,9	*(asterisk)	11,4,8
6	6	K	11,2	#	3,8	>(greater than)	11,5,8
7	7	L	11,3	@	4,8	<(less than)	11,6,8
8	8	M	11,4	(	5,8	↑	11,7,8
9	9	N	11,5	)	6,8	£	0,2,8
&	10	O	11,6	]	7,8	, (comma)	0,3,8
− (minus)	11	P	11,7	" (quotes)	11,0	%	0,4,8
A	10,1	Q	11,8	/	0,1	?	0,5,8
B	10,2	R	11,9	+	10,2,8	=	0,6,8
C	10,3	S	0,2	. (stop)	10,3,8	←	0,7,8
D	10,4	T	0,3	;	10,4,8	Space	None

48-Character code

Character	Punching	Character	Punching	Character	Punching	Character	Punching
Zero	0	A	10,1	M	11,4	Y	0,8
1	1	B	10,2	N	11,5	Z	0,9
2	2	C	10,3	O	11,6	%	1,2
3	3	D	10,4	P	11,7	¼	1,3
4	4	E	10,5	Q	11,8	−	1,4
5	5	F	10,6	R	11,9	/	1,5
6	6	G	10,7	S	0,2	½	1,6
7	7	H	10,8	T	0,3	.	1,7
8	8	I	10,9	U	0,4	@	1,8
9	9	J	11,1	V	0,5	¾	1,9
10	10	K	11,2	W	0,6	&	0,1
11	11	L	11,3	X	0,7	Space	None

FIG. 5.2. Examples of punched-card codes.

which automatically picks cards singly from the hopper and feeds them past a punching station to an output stacker. The punching station is activated by electrical signals received from the keyboard, and each column is punched one by one as details are entered by the operator. The card is automatically ejected when column 80 is punched, or sooner if so desired by the operator.

DATA COLLECTION AND PREPARATION

Keyboards are of two types: a numeric keyboard, which is operated with one hand and which requires two or more keys to be pressed simultaneously when alphabetic characters or symbols are punched; and an alphabetic keyboard, which has a layout resembling a typewriter keyboard.

A number of automatic functions can be programmed using a plugboard and special function keys on the keyboard; these include automatic spacing or skipping over nominated columns, automatic gangpunching of common data, and automatic feeding and ejecting of cards.

The original concept of punched-card data processing systems was to use one card for each transaction; a card might therefore represent a line on an invoice, a job ticket, a stock balance, a price card, an entry on a time sheet, and so on.

This concept is retained in many computer systems that immediately replace punched-card systems, but is not usually retained as a permanent feature. A computer has much greater power to edit and arrange data, and therefore it is possible to arrange for several transactions to appear on a single card, thus minimizing the time required to read cards into the computer.

Verifying Cards

There is always the possibility that a punch operator will make mistakes in reading data from the original documents or in operating the keyboard of the card punch. For this reason a second operation known as verifying is necessary. Punched card verifying machines are similar in concept to card punches except that a reading station is employed instead of a punching station. The verifier operator loads the previously punched cards into the hopper, and then uses the original documents one by one in the same sequence as the original operator. Keystrokes are repeated, and the verifying machine checks details entered by the operator against the data recorded in the cards. If the machine reveals a discrepancy a signal lamp glows and the keyboard becomes inoperable.

The verifier operator then checks the original document to see if she has made an error or whether the original card is incorrect. Incorrect cards are removed to be punched and verified once more. Usually cards which have been correctly verified are notched automatically on one edge.

Speeds of Card Preparation

The preparation of punched cards is done in batches – the cards and the original documents have to be kept in sequence within batches, and all batches must be checked for completeness at various stages before releasing the data for input to the computer. All this handling takes time and effort, and must be allowed for in estimating data preparation loads.

Batch sizes should be chosen to permit ease of handling and checking – this

SYSTEMS ANALYSIS IN BUSINESS

may depend upon the size and format of the original documents, but generally speaking might be in the order of 200–600 cards. Bigger batches may be cumbersome to handle if punching errors are detected, thus causing the batch to be held back until all corrections are made.

A trained punch or verifier operator can maintain speeds of up to 15,000 keystrokes per hour for long periods providing good punching documents are used. The speed will vary according to the arrangement of fields on the original documents, the repertoire of characters needed, the legibility of entries on the original documents, and the number of automatic functions employed, e.g. gangpunching and skipping. Automatic functions can be rated at the maximum speed of the device, say, 35 columns per second.

MAGNETIC TAPE ENCODERS

There are obvious advantages to be gained if input data can be transcribed directly on to magnetic tape: the operations of data preparation can be speeded up, large volumes of input data can be stored compactly on a single reel, the tape can be easily handled and transported, it can be read directly into the processor at high speed, and it can be used over and again, thus minimizing material costs. It is not surprising, therefore, that magnetic tape encoding machines are becoming increasingly popular where large batches of data preparation work arise.

Magnetic tape encoders consist simply of a keyboard, a storage unit, and a magnetic tape deck capable of reading and writing to a reel of tape. The keyboard is used to record data in much the same manner as a keyboard of a card punch, and the storage unit acts as a buffer between the keyboard and the tape deck.

Each memory position can store a single 7- or 9-bit character as used internationally in standard magnetic tape character codes. Usually 80 or 160, or more, positions are available in the storage unit – the size of the unit dictates the maximum block size that can be created on the tape.

Encoders are used to record data on magnetic tape and to verify that it has been correctly recorded – the mode of operation is determined by setting switches on a control panel. These machines can be programmed to carry out various automatic functions, e.g. duplication, skipping, and the entry of constant data.

When recording data the operator types on to the keyboard and details are stored in the storage unit; when a complete record is created in the store, the record is written to the tape. The tape is then backspaced automatically and the block just written is read once more to compare with details still held in the storage unit; thus a complete check is made to see that data has been transferred and recorded correctly.

DATA COLLECTION AND PREPARATION

When verifying a tape, information is transferred into the storage unit one block at a time and is compared with entries made manually on the keyboard. If a discrepancy is detected, the keyboard is rendered ineffective until the verifier operator has indicated the action required next. She may simply amend one character in the storage unit, or amend the field concerned because one or more characters have been omitted. The tape is automatically backspaced when an error is detected, and the block is overwritten when the verifier operator has checked and completed the block.

Using magnetic tape encoders rather than card preparation machines, considerable improvements in the throughput can be achieved on large batches. Data can be packed on reels of 1,200 feet length, at 200 or 556 or 800 bits per inch. Tape writing speeds of about 10 inches a second are attained, and the rewind speed is usually about 60 inches per second.

PAPER TAPE

Paper tape was used as a means of input and output to the earliest computers because telegraphic equipment (page printers, tape readers, and tape punches) provided useful means of handling data before the later development of specialized input–output peripherals. This medium has continued to be useful particularly with the development of telecommunications based applications. Paper tape devices have since undergone considerable development to match the speed requirements of computers.

Paper tape is usually supplied in reels about 1,000 feet in length, and 1 inch wide. Characters are represented by rows of holes punched across the tape and spaced along it at 10 rows per inch. Various punching codes can be used, and there may be 5, 6, 7, or 8 holes per row according to the system used. Most data processing applications use a 7-track or 8-track system; the standard recommended by the International Standards Organization is an 8-track system in which each character is represented by 7 data bits and 1 parity bit.

Methods for generating input data on paper tape vary considerably. There is one system which closely resembles that adopted for punching and verifying cards – it is known as the two-tape method. An original operator working from source documents uses a keyboard-operated device to punch a tape which is subsequently verified by a further operator. The verifying operation is carried out by a special machine which reads the original tape and compares it with entries made upon a keyboard by the second operator. If the comparison shows both entries to be correct, the details are automatically copied, character by character, by punching into a second tape which is the final verified version used for input to the computer. If a comparison shows disagreement between the entry made by the second operator and the original tape, the device stops and the verifier operator has to consult the original documents once again to determine the cause of the error.

This method has the same advantages as that used for the preparation of punched cards in that a second independent checking operation takes place, and errors are thus isolated. Of course, with punched cards it is easy to replace single cards that may be in error without disturbing the rest of the pack. With paper tape preparation it is usual to generate an entirely new tape at the verification stage so that extensive editing of the original tapes can be avoided. It is also necessary to have some automatic parity checking system built in to the output stage of paper tape verification in order to safeguard against malfunction of the output punch itself.

There are several other methods of paper tape preparation, but only one provides a fairly rigorous check upon the original transcription. In this system two tapes are originally produced by operators working independently upon the same source documents. The tapes are then compared automatically by a machine which generates yet another tape that is finally used as input. If discrepancies are detected during the comparison, the equipment stops and a senior operator must examine the original tapes and source documents to correct the error.

Comparing Paper Tape and Punched Cards

Devotees of punched cards or paper tape argue that one media has advantages over another. In particular applications, one method may be better suited, but in general the distinctions are not clear cut.

Paper tape probably permits greater flexibility in format; it is an ideal medium for using variable-length fields or records, and input records are not restricted to a maximum of 80 characters as appears to be the case with cards. However, card formats can be arranged to cope with these situations without too much difficulty, and it is not difficult to spread input records across several cards and to amalgamate them during editing and validation after input. Speeds of handling card and paper tape input are similar.

Paper Tape as a Byproduct

The most attractive feature of paper tape is that it can be captured as a byproduct of other operations. For example, paper tape punching devices can be attached to cash registers, typewriters, and other keyboard machines. Card punches can also be linked to other machines, but not with such convenience as paper tape mechanisms, particularly where data has to be captured at point of sale, e.g. in a shop or a warehouse. Most paper tape readers used as input to computers can be modified to accept special tape codes, and the costs may be relatively low compared to the costs of modifying a large number of recording devices in the field.

DATA COLLECTION AND PREPARATION

ELIMINATING REDUNDANT OPERATIONS

A systems analyst should exploit automatic data collection to the fullest extent. The operational problems of this type of data collection system are well worth conquering to obviate the need for traditional punching and verifying methods. Advantages may be even more significant if the computer system has to service a number of distant locations, e.g. byproduct paper tapes can be transmitted directly to a centralized computer to provide hourly or daily collection of the raw data. Data collection systems that obviate the need for coding, punching and verifying operations are much preferred. These operations are not productive but are merely intermediate steps in the whole data processing system; apart from adding to the costs of the job as a whole, these activities utilize floor space and trained data preparation staff which are relatively scarce resources. The operations also consume time and may affect the response provided by a system, as well as involve staff external to the computer procedures in unnecessary clerical functions.

Some of the devices and techniques described subsequently in this chapter are capable of being employed in a creative manner to develop efficient and streamlined data collection systems. The systems analyst has the responsibility to bring such matters to the fore and to evaluate them in relation to particular data collection problems.

CHARACTER AND MARK RECOGNITION SYSTEMS

In an effort to avoid the costs and delays of traditional data preparation methods, a range of devices has been developed which can directly interpret characters or marks printed, or made manually, on source documents. Thus the computer is directly able to accept information originated by other machines or by people.

This process is known generally as pattern recognition and includes currently two primary methods:

(a) Magnetized ink character recognition (MICR);
(b) Optical character recognition (OCR).

The second of these techniques includes also *mark reading*, which is a more artificial but often potent method of solving data collection problems.

Magnetic Ink Character Recognition

This system depends upon printing documents which include lines of information that are preprinted using magnetizable ink particles. When the documents are read by an MICR reader, the ink is magnetized automatically as the document is transported to a read head. The format of the magnetized

SYSTEMS ANALYSIS IN BUSINESS

characters is interpreted by the read head, which generates electrical signals, which in turn are analysed to determine basic characteristics that enable each character to be identified.

MICR systems have been predominantly used in banking institutions, wherein each subscriber is given a cheque book which contains cheques preprinted with a line of identifying information: i.e.

Subscriber's account code;
Cheque number;
Banking organization identity code;
Bank branch number.

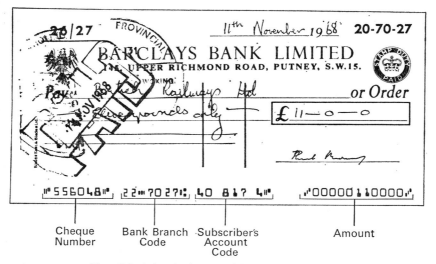

Fig. 5.3. A bank cheque with MICR characters.

An example is shown in Fig. 5.3: notice the use of special symbols to segregate the various code fields on the documents.

When a subscriber issues a cheque, he or she writes in the value of the cheque in the normal manner, and this is subsequently printed on to the cheque by a special encoder, either at the branch receiving the cheque or in some central clearing house. Thus all the data elements needed to process the transaction are recorded on the cheque, and only one of these elements is entered manually.

There are a number of machines involved in subsequent processing, but one major function is the sorting of the cheques at high speed in the clearing house to arrange the cheques into batches for each banking organization and for the various branches of each organization. This is usually achieved off-line from

DATA COLLECTION AND PREPARATION

a computer processor by a machine known as a Sorter/Reader, which reads the documents and arranges them into individual pockets according to characters detected in particular columns. Thus to sort a field containing six character positions will require the documents to be passed through the machine six times. Sorter/Readers can operate at speeds up to 2,000 documents per minute, and such machines have had a big impact upon clearing-house operations.

Other machines are capable of reading MICR documents and writing their contents to industry-compatible magnetic tape or directly into the processor of a computer system. Thus all the information on each document is captured and the system can automatically update and maintain clients' accounts, and so on.

There are two basic type founts for MICR: fount E13B, adopted as a standard in many countries for banking operations; and CMC7 which is used in France. Both systems have a range of characters including numeric digits, alphabetic characters, and some format effecting symbols, and both have a stylized fount that is not altogether easy for the human eye. In many applications a limited range of characters are used, i.e. mainly the numerals.

Some Problems of MICR

MICR systems are usually expensive to install, and the skilled maintenance of the high speed document handlers is an important factor in considering system overheads. The costs of printing are high because a close degree of precision is needed for production of lines containing MICR characters. This factor somewhat restricts the use of MICR to situations where the document is preprinted before it enters the system. Restrictions in the cost of printing and reading devices tend to restrict the format to allow one line only per document.

Encoders for adding additional fields are relatively expensive because of the small tolerance allowed in printing. This militates against their use away from centralized processing units except in cases where the application is one in the mainstream of the organizations administrative functions, i.e. a large application in which very significant reductions in costs and improvements in efficiency result.

Optical Character Recognition

OCR systems rely upon the ability of machines to distinguish the basic pattern of a character when printed on a white surface. A scanning device measures the degree of reflectance obtained when the surface of the document is illuminated, and electrical signals are generated for each character which are again analysed to generate signals representing the characters recognized.

There are less restrictions in the format of OCR readers, and the type founts

85

used are less stylized (see sample in Fig. 5.4). The International Standards Organization has developed standards for type founts which can be reproduced by many typewriters, computer line printers, tally roll printers, etc.

This flexibility has led to a more widespread use of OCR than MICR, and systems are not so restricted as to the way in which documents are initially

```
ABCDEFGH abcdefgh
IJKLMNOP ijklmnop
QRSTUVWX qrstuvwx
YZ*+,-./ yz  m  åøæ
01234567 £$:;<%>?
89          [@!#&,]
     (=)   ¨ ´ ` ^ ~ ˇ
                  ‾ ,
ÄÖÃÑÜÆØ   ↑≤≥×÷°¤
```

Note: Most OCR systems will use only a subset of this range of characters

FIG. 5.4. ECMA 'B' – an OCR type fount.

created. Some quite cheap devices have been developed to record transactions at their point of occurrence using OCR founts. For example, credit companies often issue to subscribers a simple credit card containing an account number embossed in OCR fount. The card may be presented at any one of a large number of branches to obtain goods and services on credit. A simple device known as an imprinter is available to accept the card and print the account

DATA COLLECTION AND PREPARATION

number along with other details recorded at the point of sale. Simple OCR documents generated in this way are sent daily to a data processing centre to record sales and to update customers' accounts.

In designing OCR systems the analyst should carefully evaluate the standards and tolerances for generating OCR characters, and pilot operations are advisable to test the quality of production and recognition.

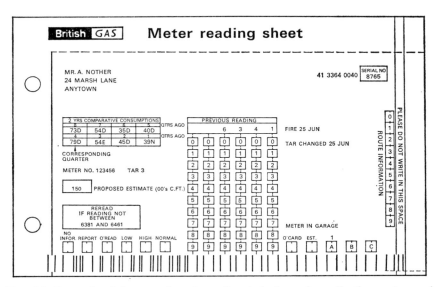

FIG. 5.5. Example of the use of mark reading techniques for collecting meter readings. (Based on an example developed by the Eastern Gas Board, and published with their permission.)

The forms themselves will generally need to be carefully preprinted so that all clock marks used by the OCR reader to register the passage of documents through the machine are positioned correctly.

OCR readers may be directly linked to a processor or to be used off-line to generate paper tape, or industry compatible magnetic tape.

Mark Reading

In many applications the data to be collected may be simply answers to questions requiring a yes/no response, e.g. in census forms. Others may require simple quantitative statements to be recorded involving a relatively small range of values or fields. These situations can be met by the use of mark reading equipment.

A mark reading device will simply scan a document and examine specific boxes to establish whether or not a mark is present. An input program examines the pattern of marks and determines their meaning in accordance with logic written into the program itself. Thus simple entries made by hand on documents are read and interpreted by the equipment without any intermediate action.

These systems are useful in surveys and for handling commercial transactions. Mark reading devices are widely used for handling meter recording in electricity- and gas-billing applications. The meter reader has to read six dials and enter details on to a meter reading slip as indicated in Fig. 5.5. There are six columns available, one per dial, and one entry has to be made in each column to record the current position of the hands on the dials.

Turn-around Systems

Turn-around systems are ones in which the basic transaction documents are generated by the computer, with key data being printed as characters or marks before the documents are released to circulate in the external procedures. As a result of handling each transaction externally, further data is recorded on the document which returns to be read by the OCR reader into the computer once more.

The significance of this system is that the key information which enables the transaction to be identified is recorded by the computer in the first place – the line printer might have to be modified to print the keys in OCR fount, but otherwise marks could be printed as dashes. The problems associated with incorrect preparation of keys are thus virtually eliminated. Gas and electricity undertakings provide examples of this type of application:

(a) Turn-around documents are generated when meters are due to be read; the meter readers visit consumers, record details of the current readings, and return them to the system;
(b) The system examines the current meter position, relates it to the position previously paid for, and generates a second turn-around document – the bill for the customer;
(c) The bill contains a detachable stub which includes payment due as well as the keys; this stub is returned to the system when payment is made and the subscribers account is updated.

The principles outlined above can be extended to cover many other applications.

Operational Problems with OCR and Mark Reading

Most optical scanning systems are expected to consume large volumes of input transactions at high speed. In many cases they are specifically employed to carry out work that would otherwise be done by many operators using

DATA COLLECTION AND PREPARATION

conventional data preparation and verification methods, thus saving the cost of labour, machines, and overheads associated with this activity. Perhaps most important they are intended to reduce the time taken to receive transactions and get them into the computer for processing. In practice, these aims are being satisfied in many organizations today, but it is worth remarking upon some of the problems that must be overcome before such plans reach fulfillment.

In most instances it is necessary to develop special input programs for handling input documents and validating the information generated by the character recognition device. These routines are usually related to the logic of the data itself and are necessary to provide additional checks upon doubtful characters or false marks encountered by the device. Applications may, therefore, initially require extra time in planning these routines and in monitoring and maintenance of them. Of course, as experience grows within an organization, each new application can be tackled with more understanding of these problems.

There are also technical problems associated with the device itself; maintenance is often more expensive, particularly for high speed document handling mechanisms, and failure of the mechanism can disrupt critical input schedules. There is, of course, the possibility of having two machines, but economic justification will be difficult unless the volumes are very large.

One area of difficulty concerns the paper used for the documents. Most machines are capable of handling a variety of form sizes, but machines are often sensitive to the thickness and quality of paper, and the opacity of the paper will be important for optical systems.

The handling of documents external to the computer room is also important. The paper should generally be resistant to creasing, but effort will need to be spent indoctrinating those who handle documents to obviate unnecessary folding, stapling, marking, clipping, or staining of forms. Where the organization has little control over those who handle documents, e.g. in public utility billing, then considerable thought is needed to determine the optimum size and format.

To make the best use of an automatic recognition device one also has to give thought to loading and operating it. High speed feeding systems usually have to work consistently for long periods of time, and the economic justification for the system may be compromised unless work flow is organized to ensure that the machine is fully loaded. It may also be necessary to have more than one operator – one feeding the machine, another unloading it, and perhaps another dealing with rejects.

All these points indicate a careful analysis of the data collection problem and perhaps a survey of the performance achieved by other users in similar situations. A pilot scheme is advisable before committing one's organization to adopt such techniques, and the implementation of the system should probably be phased to isolate problems of the type mentioned above.

ON-LINE INPUT DEVICES

The future in data collection systems will undoubtedly move in the direction of direct on-line input to the computer file concerned so that the person responsible for making the transaction does it in co-operation with the computer. All data relevant to the transaction is presented to the computer which will also select from its backing store details to assist in conducting the transaction.

In this situation, the user terminal needs a keyboard which can be used to insert data, and a printer or graphic display unit to display comments or confirmation that the transaction is acceptable.

Systems of this sort can be developed where there are a number of users each of whom may wish to order or reserve items or resources recorded in a central file system. Airline seat booking and theatre ticket reservation are applications in which this approach has been very successfully adopted. The principles can be extended to cover any application in which demands of one sort or another are assigned items from an inventory.

Types of devices suitable for such applications include teleprinters, on-line typewriters, and cathode ray tube visual display units. These devices will operate under control of a master program which will accept messages from them (i.e. inquiries, bookings, and cancellations) and which will retrieve data immediately from a direct access file system to ascertain:

(a) Whether the transaction is for a valid resource;
(b) Whether the transaction is logically correct in other respects (e.g. price, unit of issue);
(c) Whether the resource is available to allocate and if not, the availability date;
(d) What alternative resources might be used to satisfy this request.

Answers to the above questions can be displayed, in accordance with a formal procedure understood by all users, and the resources can be booked or otherwise depending upon the circumstances.

The obvious advantage of this kind of system is that the user is directly in touch with the computer programs handling his request. In this way, communication can be arranged to permit a considerable degree of self-checking so that the terminal user will find it difficult to make mistakes. Every time he makes an entry some information extracted from the file system can be displayed to help substantiate the request. Consider the following procedure in a warehouse order processing operation:

DATA COLLECTION AND PREPARATION

(a) The operator receives order documentation from a customer; he keys in his staff number and the customer number and customer's order number;

(b) The computer program validates the staff number and customer number and displays the name and address of the customer as extracted from the files;

(c) The operator checks the name and address details with those on the order and then acknowledges that these are correct;

(d) The computer requests entry of the order details;

(e) The operator enters the item code of the first item, the quantity required, and the catalogue price quoted by the user on his order;

(f) The computer displays a description of the item, its unit of issue, and the price selected from the file system;

(g) The operator checks these details against the user's order and acknowledges if they are correct;

(h) The computer indicates whether the item can be met from stock or whether it has been back-ordered for later issue on receipt of further stock. It also requests the next item on the order.

The computer system itself can take care of operations thereafter, generating packing and advice notes for orders at regular intervals.

The exact procedure adopted will differ from one application to another, but here we see in outline how the procedure can be developed to isolate errors or ambiguity at the time when the order clerk has the customer's order document to hand.

The possibility of errors is significantly reduced and the delays associated with traditional punching and verifying eliminated. To operate such a system successfully requires a considerable investment in hardware, software, and operator training. The application would probably need to be of such proportions that the organization is prepared to dedicate a large part of its computer resources and staff to the operation.

THE APPROACH TO DATA COLLECTION

The methods used must be relevant to the application at hand; the cost and time spent in data collection must also be related to the importance placed upon the application by the organization. Within these parameters the analyst should always seek to avoid potential for human errors in data collection, aim at simplicity and use labour economically. Experienced control clerks and data supervisors must be free to control and manage data, and not be tied down in mundane clerical drudgery.

The response required from the system is of importance, particularly where the processing system is an intermediate step in major operational activities. For example, in production and inventory control systems, delays in data input may delay production and keep the major part of an organization's work force idle.

Certainly data collection is one of the most difficult aspects of any operational system; care in design and in education and preparation of those who will eventually run the systems is of utmost importance.

Chapter 6

Designing File Systems— Magnetic Tape

BASIC METHOD OF HANDLING MAGNETIC TAPE FILES

A Serial Medium

Magnetic tape is one of the principal backing storage media used in computer systems – that is to say, it is used to hold files of information in an organized fashion so that programs can retrieve data as required. These files are similar in concept to the reels of magnetic tape used in a conventional tape recorder, i.e. when information is required from a particular file, the operator has to load the reel concerned and a search is made through the tape from beginning to end to retrieve the information needed. A particular organization will probably require many data files, and it is not economically feasible at present to build computers with large enough internal storage capacity to retain all the file information needed; thus backing storage media such as magnetic tape are necessary.

Magnetic tape is classified as a serial storage medium, since records can be retrieved only by reading through the file from beginning to end; each record has to be examined to see whether it displays the characteristics sought. Extracting information from a file is known as reading; records extracted in this way can be placed into the computer's main memory for processing.

To insert new information into a record on a magnetic tape file, it is necessary to read records from the file and to update them by changing the values of individual fields and perhaps by extending the length of records. If some records have been increased in length it is clearly difficult to write them back on to the same position on the reel of tape without overwriting information relating to adjacent records on the file. This aspect governs the whole approach to file processing using magnetic tape. The updating of a magnetic tape file entails reading records from the file and examining them one by one in main memory. These records may be amended or not, but will then be written to another reel of tape in the same sequence as the records from the original file. Thus at least two versions of the file will exist, and one of these will be updated to include the latest amendments. This characteristic has important benefits for the security of information since it is clearly possible to repeat the most recent updating run to get a new master file should the current master get damaged or lost for any reason. We will return to the question of file security later.

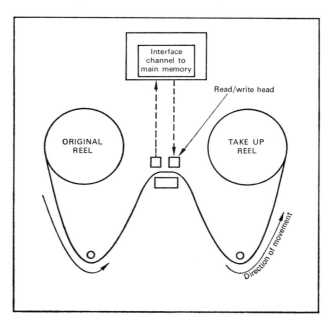

Fig. 6.1. Outline of magnetic tape deck.

In order to read and write data on magnetic tape files a device known as a tape deck is used (Fig. 6.1); a tape is wound at high speed from one reel to another past a read/write head which can either read data into memory or write data to the tape. A tape deck can be instructed by program to read data into main memory, to write data from main memory to tape, to rewind the tape at completion of a reel, and to re-position the tape to its 'beginning of tape' marker so that it is ready for reading or writing once more. Some tape systems also enable tapes to be read in reverse direction, and some permit the tape to be positioned to intermediate points along the tape according to special tape marks previously written to the tape.

Before considering methods of file processing in further detail, it is as well to consider some of the physical properties of magnetic tape and the relevance of these to the structure of individual records and blocks of data in files.

DESIGNING FILE SYSTEMS—MAGNETIC TAPE

Physical Characteristics of Magnetic Tape

Magnetic tape is usually a continuous strip of plastic about ½ inch wide and roughly 2,400 feet in length. It is coated with a magnetic oxide which can be magnetized to record information as a series of discreet bits of information which may relate to binary numbers or to numeric and alphabetic characters

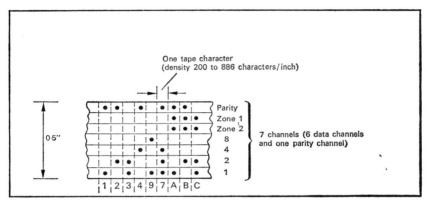

FIG. 6.2. Layout of data on 7-track ½-inch magnetic tape.

Supplier code	Location code	Supplier's Name	Location Address
1 to 6	7 to 9	10 30	31 65

Product classification codes	Discount %	Discount reason	Service rating	Total value outstanding orders	No. of orders outstanding	Total value of overdue orders	No. of overdue orders
66 75	76 to 77	78 to 79	80 to 84	85 to 88	89 to 92	93 to 96	97 to 100

FIG. 6.3. Sample record layout – a supplier directory.

represented as binary coded data. There may be usually 7 channels (there is another system using 9 channels) across the tape, 6 of which contain bits used to represent information, and the seventh is used as a parity bit to check the accuracy of read and write operations; the use of parity bits is explained further below. A set of 7 bits is referred to as a tape character (Fig. 6.2.)

A record on a magnetic tape file consists of a number of sequential tape characters which are grouped to form the required units of data. Each record may contain several fields, and one or more of these fields is used to contain key information to identify each record. For example, in a personnel file the key field might be the staff number or the name and full initials of the individual concerned. Quite obviously, the former method will avoid potential ambiguities provided sufficient control is exercised in the issue and control of staff numbers. A sample record layout for a suppliers index file is given in Fig. 6.3: here there are two keys – the main supplier number and the supplier's location code.

The capacity of a reel of magnetic tape is governed initially by the density of the recorded signals which may vary from about 200 characters per inch to 800 characters per inch according to the tape decks used – the higher the density the higher the capital cost of the deck.

Similarly, one finds variations in the operating speeds of decks from, say, 30 inches per second to 200 inches per second. At a speed of 75 inches per second and with a packing density of 556 characters per inch, about 42,000 characters can be transferred between the main processor of a computer and a tape deck in one second.

Records and Blocks

We have described a record as being a unit of information consisting of a number of related fields; each record is identified by keys which are fields containing values unique to particular records. It would be possible, and indeed sometimes is practicable, to read and write records to tape individually as separate blocks.

A block is the basic unit of transfer between a processor and a tape deck, and is identified physically on the tape by an interblock gap. When a read operation takes place, the tape starts from an interblock gap and reads characters into the main processor memory until a further interblock gap is detected.

These interblock gaps are usually about 0·56 inches or 0·75 inches in length, and they therefore consume time during reading and writing operations. They also utilize valuable space on the tape itself, and if there are too many gaps they can have a considerable effect on the efficiency of a run. This relationship is demonstrated by the timing chart shown in Fig. 6.4.

If records in a particular file are relatively short it will be far better to pack several records into a block and transfer information to and from the tape decks as large blocks. There are obviously limitations to the permitted size of blocks; if they are very large they may occupy too much space in core store when transferred to the main processor. Also tape handling operations are conducted using utility software provided by the computer manufacturer for general application; such software may impose an upper limit upon block size, and permit the user a choice up to that limit. In the final event the analyst must specify block size, bearing in mind the relevant physical constraints and the characteristics of his programs and files. More will be said about this later in the chapter.

A block may therefore consist of one record or a group of records, and if each record is of fixed size few problems are likely to be encountered. However, sometimes the record structure is such that poor tape utilization will result unless records of variable length are used.

Block size	Read/write time a million words (minutes)	Rewind time a million words (minutes)	Reel capacity (million words)	Time to read/ write block (milliseconds)
10	17·50	6·85	·35	10·5
15	12·00	4·77	·50	10·8
20	9·25	3·72	·64	11·1
25	7·60	3·10	·77	11·4
30	6·50	2·68	·89	11·7
40	5·12	2·16	1·11	12·3
50	4·30	1·85	1·30	12·9
75	3·20	1·43	1·67	14·4
100	2·65	1·22	1·96	15·9
125	2·32	1·10	2·18	17·4
150	2·10	1·02	2·36	18·9
200	1·82	·91	2·63	21·9
250	1·66	·85	2·83	24·9
300	1·55	·81	2·97	27·9
400	1·41	·76	3·17	33·9
500	1·33	·72	3·31	39·9
600	1·27	·70	3·41	45·9
700	1·23	·69	3·48	51·9
850	1·19	·67	3·56	60·9
1000	1·16	·66	3·62	69·9

Note: Packing density = 556 characters an inch. Interblock gap = 0·75 inches.

FIG. 6.4. Magnetic tape utilization and timing details.

Variable and Fixed-length Records

In some cases the nature of the data will dictate that not all fields are present in every record. If a fixed-length record format is chosen there will be redundant fields in some records resulting in poor tape utilization and a corresponding increase in the read/write time for the file concerned. The housekeeping software provided by manufacturers to control magnetic tape file handling usually permits the use of variable-length records; programs are able to determine record sizes either by the use of a word count stored at the beginning of each record or by the ability to recognize special characters used as

SYSTEMS ANALYSIS IN BUSINESS

end of record markers. With these facilities available it is up to the analyst to decide whether it is worth using them in particular situations.

Using variable-length records, the analyst may be able to reduce input/output time but will perhaps require more complex program logic to select

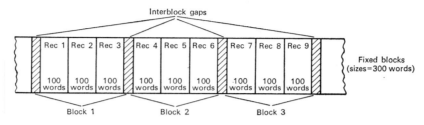

(a) Fixed-length blocks with fixed-length records

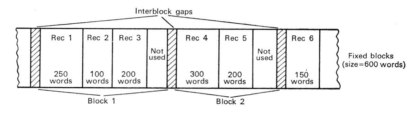

(b) Fixed-length blocks with variable-length records

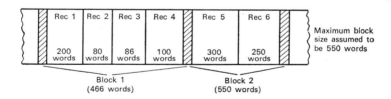

(c) Variable-length blocks containing variable-length records

FIG. 6.5. Fixed and variable-length blocks.

individual records and fields within records when they are stored in the main memory of the computer.

Some types of computers are known as character oriented, i.e. they are designed to handle fields of variable length in which redundant zeros or spaces are not used – the end of each operand being identified by a special end-of-field character. Thus although each record may contain a fixed number of fields, the overall record length may vary according to the size of the separate operands in the constituent fields of the record. For example, in a name and address file, the name A. C. SMITH LTD. will require less characters than the name BREWHOUSE, JONES AND PARTNERS.

Variable-length records may also arise in computers that operate with a storage system that is word oriented; here the situation arises not because variable-length operands are used but because the records can consist of a number of subrecords which need not all be present in every case.

Where variable records are used it is usually necessary to have variable block sizes. It is not desirable to split individual records across blocks since this presents unwarranted processing complexities, and it is still necessary to operate within some predetermined maximum block size; this size having been specified according to the storage capacity available for input and output buffers in the program concerned. The various conditions that can be catered for are summarized in Fig. 6.5.

Variable-field lengths, of course, can only be handled by character-oriented systems, but variable record sizes and variable block sizes can be used at the discretion of the analyst according to the nature of the data to be handled. The advantages and disadvantages have to be considered for particular situations, and relate to the degree of variability likely to be encountered for a given volume of records and the frequency of file processing operations. If savings in input/output time outweigh the implications for programming complexity, then variable-length working is justified.

BASIC RUN TYPES

Magnetic tape units operate at faster speeds than slow speed peripherals such as card readers, line printers, and paper tape input/output devices. Computer systems which rely largely on magnetic tape storage are generally more efficient if runs are designed to minimize the mixing of slow speed peripherals and magnetic tape decks in the same runs. This ideal is not always possible. However, in general, magnetic tape processing systems incorporate the following run types:

(1) Input transcription;
(2) Validation;
(3) Editing;
(4) Sorting;
(5) Updating;
(6) Reporting.

A run-flowchart showing these run types in a simple batch processing system is given in Fig. 6.6.

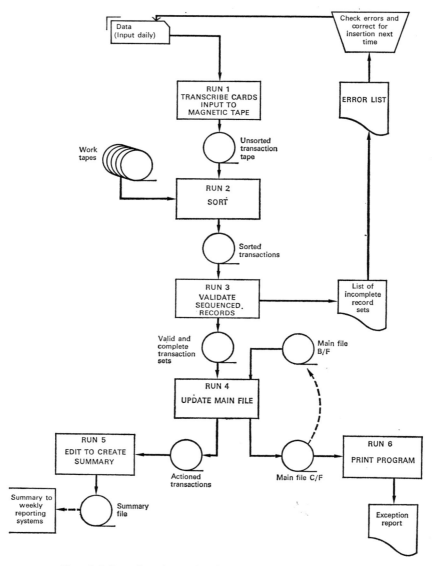

FIG. 6.6. Run-flowchart of a daily batch processing system.

DESIGNING FILE SYSTEMS—MAGNETIC TAPE

Run 1. Input Transcription

The initial data fed to a computer on, say, cards or paper tape, has to be converted to a high speed backing store media. It is usual to achieve all conversion in an initial input run so that subsequent processing will entail the use of high speed peripherals only. In run 1 of the example, cards are transferred directly to magnetic tape. Some cards may not be in correct format or may contain fields which present illogical data conditions; these records are usually detected by the input program and are not written to tape but are listed for manual correction and re-entry at a later stage.

Run 2. Sorting

One of the most significant characteristics of any data processing system is the ability to rearrange records at high speed in order to achieve a sequence suitable for producing given analyses or for updating a main file.

To update a main file it is necessary to match the keys of transaction records with their relevant master records so that transactions can be used to amend fields in master records. This matching operation requires the master and the transaction files to be ordered in a prearranged sequence.

Run 3. Validation

Validation entails the checking of input records to see that they conform to certain standards of accuracy needed in a particular system. This may include logical checking of conditions exhibited in certain fields of the input records or checking of data conditions between different records. Often, validation at the record level takes place during input transcription as described above. However, validation is shown separately in run 3 of the example because it is assumed in this case that there should be four transaction types for each amendment, and sequence checking is needed after the sort to make sure that the transaction file contains complete sets of transactions.

Run 4. File Updating

Updating is the process whereby a main file (e.g. an inventory file, or a staff records file, etc.) is amended to reflect the latest state of affairs. Records may be added to the file or be deleted, and amendments may be made to fields of main file records. There may be several transactions to be applied to some of the master records, but perhaps only a small percentage of the master records will be amended in the same run. In Fig. 6.6 updating takes place in run 4; the new master file thus created is carried forward for inclusion in the next updating cycle.

Run 5. Editing

Editing takes place when it is necessary to apply amendments to all or some of the records in a file or when it is necessary to create a subfile from another. In the example, the transactions are summarized in run 5 after main file updating, and it is assumed in this case that the resultant file is for use in some other part of the system.

Run 6. Reporting

The end result of a system is usually to provide output of one sort or another which will either be a document having some functional purpose, such as an invoice or a pay slip, or will be a report giving information for control purposes. In Fig. 6.6 the main file is examined in run 6 after updating has been completed, to identify and display certain records which have been amended during the run and which, as a result, exhibit characteristics warranting action by management.

SECURITY OF DATA

Naturally, when an organization decides to keep its vital records in a computer file it will wish to be confident that information cannot be lost or damaged. Magnetic tape systems have many inbuilt checks which serve to monitor the accuracy of file processing operations.

Every time information is transmitted between a tape deck and a central processor, a self-checking system is operative to detect errors caused by distortion of signals in input/output operations. If individual bits are lost in a transfer, an attempt is made to read the block once more; if errors are persistent, the program is stopped and an error signal is given to the operator. This bit-checking system is called parity checking: with an *even parity* check each tape character is checked to see that it has an even number of bits present (see Fig. 6.2). In addition, each of the seven tape channels would be checked to see that an even number of bits are present for the block concerned. Parity is achieved by deliberately adding a parity bit to each character and a parity character to each block; these elements are always present but are not counted as part of the data. Some parity systems use *odd parity*, but the principles are the same.

Software systems used to handle magnetic tape files conduct checks upon information contained in header labels at the front of each file. These checks are designed to prevent a program from accessing incorrect files or overwriting vital information on another file. Header labels include the following:

(a) Name of the file – usually about 12 or 16 characters in length and assigned by the user to identify this particular file of data;
(b) The date this version of the file was written to the tape;

DESIGNING FILE SYSTEMS—MAGNETIC TAPE

(c) The date at which the information in the file can be overwritten and the reel released for another purpose;

(d) The reel number – if the file is a large one occupying more than one reel of tape for each generation, the continuation reels are numbered 2, 3, 4, etc.;

(e) Generation number – a number which indicates the relevance of any version of a file to the current master; all generation numbers are logged in the tape library, and from these manual records it is possible to check the importance of any particular generation. Normally, three generations of a master file are retained to facilitate recovery from error conditions or possible tape damage.

Master files must never be overwritten until enough generations exist to safeguard against failure. Sometimes mechanical devices are fitted to the reels themselves, which prevent the reels from being opened as output tapes. These devices are known as *write inhibit rings*; sometimes an opposite convention is used and *write permit rings* are necessary for any file required for use as output.

BATCH PROCESSING

The system shown in Fig. 6.6 requires several preparatory runs to be undertaken before updating or reporting can take place. The runs 1, 2 and 3 are a prerequisite of the updating, but in themselves are not really productive. This is typical of a magnetic tape file system – it is never practicable to deal with transactions singly or directly as they arise, but rather it is necessary to accumulate and batch transactions until a worthwhile volume of data is ready for processing.

It should be noted that in every updating cycle the master file has to be read from beginning to end and each record of the file examined for matching with the transactions. It would probably be uneconomic to run a complete updating of a master file containing 100,000 records if only 200 transactions were on hand. One cannot be dogmatic because it depends on the nature of the system and the use made of the output, but generally the processing schedule will be arranged so that larger transaction batches are processed.

Batch processing systems impose regular schedules for the collection and preparation of input data and thus provide a routine around which the clerical and supporting functions can be organized. Processing can take place daily, twice daily, weekly, monthly, or whatever, according to the dictates of the application. In the following pages we now consider the different types of file processing runs in more detail.

INPUT PROGRAMS

An input routine characteristically will accept data from a slow-speed peripheral such as a card reader, and will write output to a magnetic tape file. Since the input device will operate at slower speed than the magnetic tape unit, the program can be classified as input limited, i.e. processing will be suspended from time to time during the run while awaiting data from the input device. This time can be conveniently utilized for processing the data read from the cards so that editing operations or validation checks can be performed in parallel with card reading.

It is important also to check the completeness of the data by generating control figures for data entering the system. These control figures are then compared with totals generated manually during data preparation, and are also used to provide check figures for validating the operation of programs in subsequent stages of the system.

It is usual to divide input data into a number of batches, e.g. a batch being a pack of cards or a reel of paper tape. At the end of each input batch there may be a record containing a batch identifier and control figures originally generated by totalling predetermined fields on the original source documents. During the input run a record count is maintained for each batch along with totals from specified fields on the input records. The record count and totals are then checked with the control information on the end of batch cards.

The control figures arising from an input run can be listed onto a line printer to provide a control record for use in vetting the accuracy of the input operation. If each batch is identified with a unique number it will be a simple matter to check that all batches have been read to tape. The record counts and the batch control totals are also checked to vet that records (e.g. input cards) have not been lost or misplaced. Input programs may also include validation and editing activities: these are dealt with below.

VALIDATION

It is desirable to check data entering a system as soon as possible so that incorrect data can be isolated before its effects are felt in file updating or reporting. We have seen above how input routines can be used to verify the completeness of data, but this still leaves the possibility that data on the source documents is incorrect or that it has been transcribed incorrectly during data preparation. Validation checks are designed to check that information contained in various fields of the input records is logically compatible or that complete groups of records representing an identifiable set of transactions are present.

Validation can be performed in a separate run which reads a new input tape and checks individual records listing invalid records on the line printer with suitable diagnostic comments; at the same time valid records are written to another output tape for processing further. The invalid records have then to be corrected and re-input once more. The nature of the system may dictate that all errors have to be corrected and re-inserted before continuing the processing cycle, but it may be possible and practicable to correct amendments later and include them in a subsequent updating cycle.

Validation checks may include checking values of characters or fields in the input records, e.g. to verify that there are no alphabetic characters in a field reserved for numeric values – fields may also be checked to see that they are within predetermined ranges, say from £1 to £10,000 – and a positive value in one field may imply that another field is always zero. Such checks are related to the nature of the data and the degree of accuracy considered necessary for a particular system. As far as possible it is better to complete validation checks in the initial transcription run where there should be spare processing time.

EDITING

Editing can take place in:

(a) Input transcription programs;
(b) Intermediate file handling;
(c) Output transcription programs.

Editing is the process whereby records are converted from one format to another. This activity is usually included in input transcription programs to convert from external data formats to those that are to be used within the computer files, e.g. conversion of numeric fields in character form to binary form.

Separate editing runs may sometimes be necessary where data is passed from one system to another. If the systems concerned have been designed in an integrated fashion, such editing would not be intended to change code structures of individual data elements, but more likely would be intended to create a summarized file of information. For example, a transaction file may contain several thousand individual records of accounting transactions which can be summarized under, say, a hundred cost centres.

Editing will also take place in output runs when data are converted from internal formats back to external formats once more.

Editing during input processing will normally take place for records that have passed the validation checks. All fields that are numeric and which are to be used in arithmetic operations at some later point in the system are generally converted to binary format immediately upon input. Sometimes fields on input records are rearranged in order to get away from restrictions that may be imposed by the input media, and it may also be necessary to expand input records to include descriptive data selected from a look-up table in the input run (see Indexing below).

A common requirement in input runs is to edit fields which are to be used as keys into a standard form that will achieve a desired sequence when sorting, e.g. preceding space characters may need to be converted to zeros in an input field which is right justified.

Sometimes data has to be added to records during input runs; this will include the insertion of transaction codes to identify types of records, transaction dates, or even transaction serial numbers. In an order processing system, for example, a unique number may have to be allocated to distinguish every order and individual line items on orders; and the date may be added so that orders can be handled in priority.

INDEXING

Indexing is really a special editing category in which some descriptive data are appended to records according to key information contained in the record. For example, invoicing records may attract name and address information according to coding contained in a customer number field in the original input records.

In magnetic tape based operations, there are two ways of achieving such indexing:

(a) By means of a look-up table in core store;
(b) By a special indexing run using a separate index file.

The first method is potentially more efficient, since if the table can be stored in main memory the input records can be indexed without first being sorted into sequence. Indeed, if the processing operations are not too extensive the indexing can take place along with the validation and general editing needed in the initial input program. In core store there would need to be a table of entries quoting the various key data elements along with their descriptive entries. In the invoicing example mentioned above, the key would be the customer number and the entries the name and address details.

However, if the number of possible entries in the table is large, the input program might occupy too much space in main memory, and therefore searches through the table might be so protracted as to make the run processor limited rather than limited by the slow speed of the input peripheral.

If a separate index file is used, the transactions must first be sorted to the same sequence as the index file, and the two files are then input to an editing program which will match records and copy across details from the master index to the transaction records. The resultant output file may not then be in

DESIGNING FILE SYSTEMS—MAGNETIC TAPE

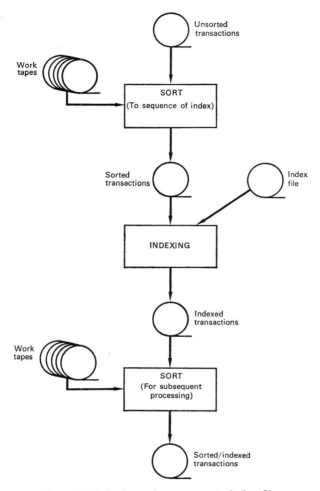

FIG. 6.7. Indexing using a separate index file.

an appropriate sequence for the next processing stage, and the indexed transaction file may need to be sorted once more as shown in Fig. 6.7.

SORTING

As we have seen in previous examples, sorting figures prominently in magnetic tape file processing. Sorts give few problems to programmers and analysts since they can be performed using standard software provided by the computer manufacturer; all the user has to do is provide parameters to describe the keys used in his file and the sequence desired.

However, although sorts are powerful tools in achieving file processing, they are in themselves not productive operations since they do no more than arrange a file in sequence. They may occupy the machine for lengthy periods and may require several tape decks – the more decks allocated the faster the sort. The analyst should be watchful in designing a system to avoid unnecessary sorting operations. Indeed, this consideration will be one of the main factors in deciding how many main files to maintain and what the relationships are between one file and another. We have already seen in the discussion of indexing how easily additional sorting operations are incurred.

As a general rule it is better to sort transaction files rather than main files, since at any moment in time a transaction file is likely to contain fewer and smaller records. Extending this principle, it is also better to sort subfiles used for output rather than complete main files. Thus in magnetic tape processing the following form useful guiding principles:

(a) Sort transaction data entering the system into an appropriate sequence for updating master files;
(b) Update and maintain master files in one preferred sequence and avoid unnecessary sorts;
(c) Create subfiles of exception records from main files and sort them to the desired sequence for reporting operations.

When we deal with file maintenance systems below, the treatment of transactions changing keys on main files is considered. Again, it will be observed that the run structure is aimed at obviating the need to sort the main file.

Sometimes it is worth incorporating editing operations in sorts – the desired processing is undertaken during the initial input pass or the final output pass of the sort operation. Sort generator programs supplied by manufacturers often provide facilities for incorporating own-coding to meet this requirement.

To summarize: sorting operations are simply intermediate stages in a file processing system – speed is governed by the number of records in a file, the size of the individual records, the number of tape decks used, the number of hardware input/output channels available to service the decks, and the operating speed of the tape deck. The size of the keys, the number of keys, and their position within the record structure of the file concerned are not very significant factors in determining run time.

UPDATING AND MAINTAINING MAIN FILES

Main Files and their Contents

Main files are used to hold a permanent record of essential information needed in the operation of a business enterprise. Ideally, a computer should

DESIGNING FILE SYSTEMS—MAGNETIC TAPE

be able to store all necessary information in a coordinated fashion so that documentation can be generated to support operational functions, as well as provide routine and *ad hoc* information to managers and other users. The amount of information required to satisfy these overall requirements for any but the smallest organizations will far exceed the capacity of a single reel of magnetic tape. Therefore information is classified into related categories and a number of major files are established and maintained to satisfy various functional needs.

Each file consists of a large number of individual records which are arranged in sequence according to certain fields which are nominated as keys. The following examples indicate the types of files that may be encountered in simple cases:

File	Key
Personnel file	Employee number
Stock file	Product code
Customer file	Customer code

Some main files will have more keys than this; e.g. a stock file for an inventory deployed over a number of locations might have as keys depot number, stock account code, part number. The analyst will then have to decide which are the primary keys in sequencing the file: e.g.

Part number, within stock account, within depot;
Depot, within part number, within stock account;
Part number, within depot, within stock account.

The choice will depend upon the application, and there are probably two points that are considered first:

(a) Can the data collection be arranged to capture the key data for each transaction?
(b) What output is desired from the system and what is the best sequence to deliver the output?

In the example given above it may not be easy to capture the stock account code for all transactions that effect depot stocks. It is unlikely, therefore, that this would be a primary key: indeed, it will probably be used as a key only if it can be captured in an indexing run against a catalogue file prior to the stock file updating run.

The depot number and the part number have probably equal claims, but if the updating operation is intended to produce exception reports sequenced by depot number it is quite likely that this will be the major key. However, if a file is intended to satisfy many functions and applications, the output desired will be in varying formats and sequences and the considerations become more complex.

The simple depot stock file may be required to satisfy the following functions:

(a) Provide stock records to cover stock audit requirements;
(b) Provide a basis for financial accounting and generate information for determining commercial policy;
(c) Provide information for forecasting and provisioning control at each location;
(d) Generate information for making obsolescence assessments.

	Format	Characters	Words	
Main Record				
Part No.	BCD	8	2	
Description	BCD	20	5	
Price	Binary		1	
Stock account code	BCD	4	1	
Supplier code	BCD	6	2	
Base stock quantity	Binary		1	
Overall value base stock	Binary		1	
On order from supplier	Binary		1	
Demands-on-hand from depots	Binary		1	
			15	
Sub-Record				
Depot No.	BCD	2	1	⎫
Stock at depot	Binary		1	Repeated
Minimum stock level	Binary		1	up to
Maximum stock level	Binary		1	20 times,
Average monthly demand	Binary		1	i.e. once
On order from base	Binary		1	per depot
Age of demand history	Binary		1	⎭
			7	

Note: Record size will be a minimum of 15 words, but may be variable in increments of 7 words up to a maximum 155 words.

FIG. 6.8. Arrangement of data elements in a simple depot stock records file.

The data elements needed to meet these requirements have to be stored within records in the file system. A set of data elements to cover the functions given above is provided in Fig. 6.8.

It is apparent from this list that much of the data is relevant to each part number and not to the location at which parts are stocked. In this case a variable length record has been chosen to embrace all data relevant to a part

DESIGNING FILE SYSTEMS—MAGNETIC TAPE

number, with information for each depot in a separate sub-record: the file is maintained in part number sequence.

An alternative solution might have been to keep two main files: a catalogue file in part number sequence, and a depot stock file in part number sequence within depot. In the final analysis the best sequence is the one which produces desired results but minimizes the storage capacity required, and minimizes the amount of machine time necessary in processing. These things have to be considered one against another, and several solutions have to be evaluated before committing oneself to a particular design.

Data in Main Files

Some data elements are relatively fixed once they are committed to the file; facilities have to be provided to allow changes to be made to these elements, but their nature is such that changes are infrequent. These will include descriptive fields such as item descriptions, names, addresses, national insurance numbers. In many instances facilities to amend fields are necessary just to recover from situations resulting from errors in coding or data preparation.

Other fields are changing constantly, and routines may have to be provided to update them hourly, daily, or weekly as the case may be, e.g. labour hours, stock levels, machine usage, sales achievement statistics, seat bookings, bank balance and so on.

The nature of the data may be such that certain data elements on the main file must be updated before other transaction types can be applied to the file. This may entail having two updating runs, or more likely require the transactions to be coded so that they can be appropriately sequenced before carrying out a single file updating operation.

Simple Updating Methods

The basic operations carried out in a main file maintenance program will include:

Deleting records;
Adding records;
Amending fields in records.

The mode of operation for an updating program is to read records from the main file into memory and to read records from the transaction file and apply them one after another to the corresponding main file record. When all the transactions relevant to a main record have been actioned, it is written to the carried forward main file, and the next master record from the brought-forward file is dealt with. Those main file records, which are not matched with transactions, are simply copied directly to the carried-forward file.

A run-flowchart depicting these operations is shown in Fig. 6.9. Here three

SYSTEMS ANALYSIS IN BUSINESS

output tapes are produced: a carried forward main file; a changes file consisting of transactions which have been actioned; and an error file, e.g. of unmatched transactions.

The error file will be listed and the transactions checked for inaccuracies, perhaps after being first sorted to a desired sequence. The simplest errors will be those where the keys have not matched a main file record. The main file may be incomplete or perhaps the keys of the transactions concerned have been incorrectly coded. This type of problem is dealt with by the data control department who will either arrange to correct the main file or the transaction

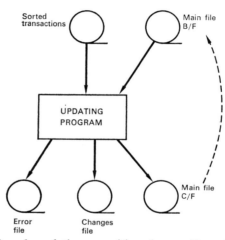

FIG. 6.9. Sample updating run with a changes file and error file.

records. Some errors may be more complex and result from illogical or incompatible relationships between data elements on the two input files. At the end of an updating run it may be necesssary to report upon exception conditions exhibited by main file records, e.g. stock levels below minimum. In this example the main file would have to be read once more and a subfile created to report these events. The actioned transactions on the changes file can also be analysed in preparation for printing or used for entry to another subsystem. In some situations it may be necessary to repeat the updating cycle after correcting all errors before performing these subsequent activities; in many cases, if the volume of errors is small, corrections can be made on the next updating cycle.

Re-identification of Master Records

As far as possible one attempts to design all master files so that records have a unique identity throughout their life within the file. However, in practice this ideal cannot always be attained, and it becomes necessary to amend the key fields of some records from time to time.

The simplest way of doing this is to post transactions to the master file which will amend the keys of relevant records, and then sort the main file before using it once more. However, if the file is a large one, this can be very

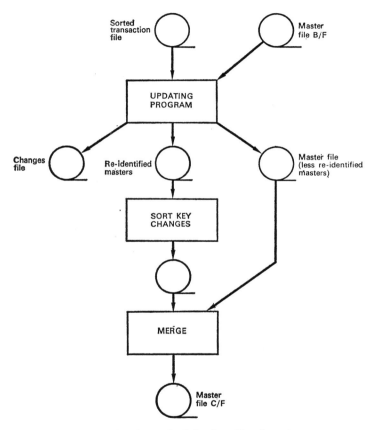

FIG. 6.10. Simple method for handling key changes.

wasteful of machine time, particularly since there will be relatively few key changes at any one time in a well-designed system.

A method of minimizing the time spent in sorting is demonstrated in Fig. 6.10. Here, during a normal updating cycle, some main file records are re-identified and output to a separate key changes tape and not to the carried forward main file. The key change file is then sorted and merged once more into the main file. Although this does eliminate the need to sort the main file, it is still necessary to read right through the main file once more to complete the merging operation.

Depending upon the nature of the system it may be possible to retain the key changed master records on a separate output file until the next processing cycle as shown in Fig. 6.11. This method has the disadvantage that the main file will be incomplete during any period in which sequence changes are being

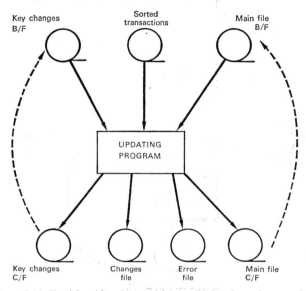

FIG. 6.11. Re-identifications with circulating key changes file.

actioned – the sequence changes from one cycle are not re-input until the main file is next updated. Whichever of these two methods is used, the data collection problem remains: How can amendments be controlled so that they quote current keys only? If old keys are used they will not be matched with a main file record and will have to be listed as errors and corrected.

A third method can be adopted which improves control, since it entails retaining a stub record of all previously re-identified keys on the main file. The stub records will appear in the appropriate sequence on the file, but will contain simply a reference to the new keys under which heading the current information is stored. If a transaction record enters the system quoting previously superceded keys, the latest key details stored in the stub record are used to overwrite the keys in the transaction record. The changed transactions are output to a separate file and can then be reintroduced to further update the main file on a subsequent run as shown in Fig. 6.12.

DESIGNING FILE SYSTEMS—MAGNETIC TAPE

Adding and Deleting Records

From the programming aspect the addition of a new record or the deletion of an existing record presents few problems to the systems analyst, but it is essential that the data collection routines are organized so that additions and

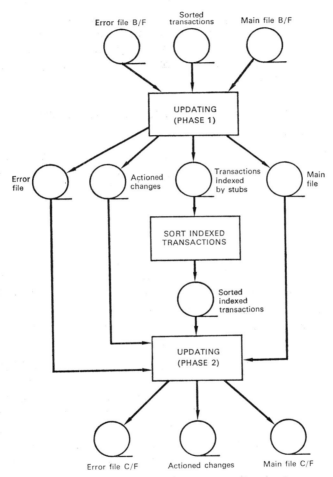

FIG. 6.12. Re-identification using stub records.

deletions are co-ordinated with other types of amendments. An updating program might be arranged to deal with transactions in the following sequence, against the possibility of needing to apply all the transaction types to one master record in the same run:

115

(a) Additions;
(b) Amendments to data fields;
(c) Amendments to key fields;
(d) Deletions.

Generally speaking, a record will never be deleted unless there is certainty that there will not be cause to update it or access it in any way. However, if old records are stripped from the main file on a routine basis, there may be considerable savings in processing time and in magnetic tape utilization. Perhaps the best solution is to retain obsolescent records on a history file where they can be analysed if necessary.

REPORTING FROM MAGNETIC TAPE FILES

It is possible to classify reporting as:

(a) Listing error conditions along with diagnostic information for the data control section or line departments to correct;
(b) Production of functional documentation or information summaries for management.

Error Reporting

Error situations generally arise where there is difficulty in matching keys of a transaction record against keys of a corresponding master record, or where there is some illogical condition between some data elements on the main file record and a corresponding transaction record. The possibilities are:

(a) The master record is not on the file;
(b) The master record has been incorrectly coded or maintained;
(c) The transaction record has been incorrectly coded and recorded.

To find out what has gone wrong, details of the transaction will have to be printed on to a line printer for visual scrutiny by the data control staff. Not only must the keys be printed but also some data relevant to the error condition, including perhaps the number of the original source document. This information is needed to help diagnose the fault and locate the original source document; if necessary, the batch control number used in data preparation should be carried around in transaction records and printed in error reports to speed identification and selection of source documents.

Choice of Output Media for Error Reports

Where errors stem from a file updating operation, the analyst will naturally try to minimize output time for the error records, and may therefore choose to write them to a magnetic tape file. However, the volume of errors should be relatively small in a good system, and it may therefore be wasteful to allocate a tape deck for this purpose especially in a multi-programming environment. A line printer is an obvious candidate for this job since it directly provides a hard copy print-out; however, all programs are potential users of a line printer, but a job may be easier to schedule in a multi-programming system if it does not call for a line printer. The use of other slow speed devices should therefore be considered: a paper tape punch provides a good medium for low volume output, and the paper tape file can be read later to list error records whenever there is time available on a line printer.

As far as possible, a system should be designed to retain error records within the computer files until effective action is taken by the user to correct them. An error file can be circulated within an updating system, and to clear records from this error file it will be necessary to input amending transactions to correct either the main file record or the error transaction itself.

Reporting from Data Files

To print details from a magnetic tape file it is necessary to read data blocks from the tape and examine them to see if the records contained therein are relevant to the report, and if so to list them directly to the line printer. Control totals and summaries required in the output can be accumulated in main memory and printed out at relevant points, e.g. upon changes in control data and/or at the end of the run.

As a general rule, however, particular reports required will not be in the sequence of the main file, and may relate to a subset of the total records only. In this situation it will probably be better to create a subfile in print format, to sort this subfile to the sequence required in the output report, and then list the records. This method will obviate the need to sort the main file and will minimize the amount of processing needed in the final output run; all output editing can be achieved in the initial run which creates the print file.

Output programs are likely to be altered frequently as line management and other users improve operations within their departments. It is preferable, therefore, to keep all output procedures as simple as possible, even at the expense of a little run time efficiency. Unless a report is aimed at satisfying some more or less static functional need, the analyst will be well advised to plan for flexibility in reporting routines.

Reports generated for management information should be summaries rather than long lists of detailed transactions, and, as far as practicable, reporting routines for line management should be based on the principle that users will be presented with only exception conditions requiring attention and action.

Attempts should be made to visualize the impact that the system output will have on the users operation, and later requirements should be anticipated so that they can be allowed for in the initial planning of reporting subsystems.

RESTARTS

It is advisable when designing any computer processing system to consider the steps that will be necessary to restart a job in the computer room should it be interrupted by some unscheduled event, e.g. equipment failure. Computer operations staff are likely to be very frustrated if a large job has to be re-run because of a failure in the final 5 minutes of operation. A facility to restart from a predetermined point is vital for any run which has a running time longer than about 30 minutes.

A dump and restart routine is a standard facility made available by most computer manufacturers as part of the basic utility software provided with the equipment. A dump routine can be incorporated in a program when it is first compiled. At certain points in the program the analyst can arrange that the routine is entered to dump details of the current state of the program and the status of the peripherals that it uses, to a special dump file on magnetic tape.

The dump routine will usually output a few details about the program concerned for use by the operator as parameters when the job is restarted. The restart routine is usually kept on the library tape and is activated by an operator when a program is to be restarted. The operator provides the parameters generated at the most recent dump point and the restart routine repositions the magnetic tape files to the data blocks current at the time of the dump point and regenerates in core store the internal memory conditions prevailing at that time.

Dumps and restarts are particularly important in lengthy runs which utilize slow-speed peripherals such as card readers or line printers. It would be particularly irksome and wasteful to have to reprint every page of a large output report simply because a printer fault had spoilt the last few pages. Any restart routine should enable the program to generate and maintain data for control accounts and totals generated within the program. The manufacturers' utility software should be used wherever possible to control dumps and restarts, and an understanding of the facilities available is essential.

SCHEDULING IMPLICATIONS

The analyst has to consider the likely difficulties that may arise in operating his file processing system in the computer room. The discussion above on dumps and restarts is related to this topic; we have already seen how the analyst has to foresee operational difficulties that can arise. He should observe principles which will tend to minimize the amount of operating activity needed and to simplify the problems concerned with scheduling the work.

First of all, the analyst is constrained by the equipment available to him; if the computer system is a large one it may have ample amounts of core store available and a variety of input/output devices. It is also likely that the computer will be a multi-programming system, i.e. capable of controlling several programs concurrently on a time-sharing basis. Potentially, therefore, the analyst may have a wide range of equipment resources allowing him to compress his processing into a few runs to minimize operator activity.

However, a program which uses too many peripheral units becomes difficult to schedule with other programs, and may also demand rather more than its fair share of internal memory. Some general standards concerning computer run-structures should be adopted within the systems department to avoid such difficulties. Programs can be considered in two broad categories to aid the development of such standards:

(a) Peripheral-limited programs;
(b) Processor-limited programs.

Programs which utilize slow speed input/output devices are generally peripheral-limited, i.e. the processing operations are suspended from time to time while the program is awaiting signals from the device concerned. Thus the elapsed time for a run is determined by the maximum speed of the slow speed devices used and the volume of transactions to be handled by the devices.

Processor-limited programs arise where the amount of processing done for each record is such that delays are experienced in requesting transfers to and from the input/output units.

A mix of processor-limited and peripheral-limited programs provides a good basis upon which to schedule computer work for a multi-programming system. Peripheral limited programs are usually given a high priority at run time, so that they are suspended whenever they are held up awaiting the completion of an input/output operation. When this occurs, control is automatically handed by the operating system to another program stored in the computer system. Processor-limited programs are given relatively low priority, otherwise they tend to absorb all the computer time and resources available and prevent the time-sharing facility from being effectively utilized. Thus in multi-programming systems, the emphasis in computer procedure design is shifted from maximizing efficiency within the particular job towards meeting more generalized objectives that will permit an effective realization of the computer's time-sharing capability. The following points are worthy of consideration in these circumstances:

(a) Input and output transcription programs should be reserved for converting data from one medium to another; they should not contain too much processing other than that needed to carry out relevant conversion and validation operations;

SYSTEMS ANALYSIS IN BUSINESS

(b) Updating programs should not call for the use of a large number of magnetic tape drives for lengthy periods;
(c) If slow speed peripherals are needed in updating programs, they should be reserved for low-volume input/output and be released by the program as soon as their function has been fulfilled;
(d) Updating and editing programs should not be constrained by the use of slow speed peripherals; attempts should be made to capture all forms of input data in one run and all subsequent operations, except final output, should involve magnetic tape to magnetic tape processing.

These principles may, if followed to the letter, result in program suites which consist of a number of runs of a more or less standard shape. There may be more runs than necessary in achieving a particular processing objective, but, on the other hand, individual program structures may be simplified making subsequent amendments easier to plan and implement. The analyst should try to make the task of the computer operations department less restricted, and design file processing structures which are easy to schedule regardless of the prevailing workload in the computer room.

AIMING FOR EFFICIENT FILE HANDLING

Earlier in this chapter we discussed the relative merits of using fixed- and variable-length records or blocks. It was suggested that the ultimate decision in these matters depends upon the nature of the data in the file concerned; there are usually a number of factors to be traded off in making these decisions, viz. increasing block size and packing fields closely within records improves tape utilization and reduces input/output time; but the core store capacity required is increased, and programming logic may be rendered more complex.

The sensible use of input/output buffers is the key to efficient magnetic tape file processing, particularly in editing or updating runs in which only magnetic tape input and output is used. These runs should not be processor limited, i.e. the magnetic tape drives should be operated continuously throughout the run at maximum speed. In a multi-programming environment these runs should be marginally peripheral limited so that they can time-share effectively with those more severely limited by slower speed peripherals.

In seeking to comply with the need to minimize input/output time and yet secure good utilization of main memory in a program, attempts should be made to balance the input and output operations. A magnetic tape processing run is said to be balanced if the time needed to read or write data blocks just exceeds the time needed to process a block of data; then, by using a system of double input/output buffers, processing and input/output operations can be overlapped.

DESIGNING FILE SYSTEMS–MAGNETIC TAPE

In Fig. 6.13 an arrangement of buffers is shown for dealing with a simple editing program in which one file is read and another is written. The input file has two buffers and data blocks are read alternatively into buffer A and buffer B, and output file is similarly created by writing first from buffer C and then from buffer D. Thus at any moment in time the input and output tapes are operating, and processing is taking place to receive and assemble data in an input buffer, to edit data and transfer it to an output buffer, and to write from an output buffer to the output file.

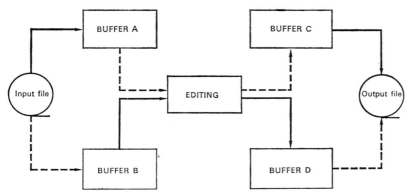

FIG. 6.13. Double buffering to time-share input and output.

To obtain balance, the analyst has to estimate the amount of processing time required to handle and edit groups of records and then establish a block size and buffer size which will result in an input/output time that marginally exceeds the processing time. If processing time for a group of records exceeds the corresponding peripheral time to read or write a buffer, then peripheral activity will be suspended until the program has completed the processing of records in the buffer.

TIMING FILE PROCESSING SYSTEMS

The systems analyst is responsible for evaluating the costs and benefits of any new system so that management can determine the opportunity costs of proceeding with the implementation. The computer time needed to operate the system is one of the prime resources that have to be included in the costs, and a standard approach is needed in making timing estimates within the systems department. The major factors considered in generating such estimates are:

(a) Estimates of the volume of transactions handled by the system;
(b) The peripheral time in each run resulting from handling these transactions;

(c) The overall elapsed time for each run considering the ability to overlap input/output activities by time-sharing;
(d) The processor time needed for arithmetic and logical operations in each run;
(e) Overall running time per run bearing in mind the ability to overlap processing and peripheral activity;
(f) The time to set up and take down the job when operating the system.

Information resulting from such estimates can be used to develop outline computer running schedules, but in practice operating times may vary according to the volumes handled in a particular run, the effects of other programs that may be processed concurrently, and the priority assigned to programs at run time.

Estimating Data Volumes

Usually it is possible to estimate the volume of transactions that may occur by observation of documentation and events arising in the departments being studied. These transactions can be classified by type according to the amount of processing necessary, and average volumes should be established for each category in a given period.

Calculating Individual Peripheral Times

Peripheral times are determined by the number and size of records handled by a particular program, and the operating speed of the devices used to handle the records. It is assumed here that good programming standards are used, that input/output routines are balanced, and input/output buffers are used to keep individual peripherals operating at maximum speed.

For slow speed input/output devices (e.g. card readers, paper tape readers, line printers, etc.) the analyst determines the volumes handled and simply calculates peripheral time by reference to the rated speed of the device.

Magnetic tape processing times are evaluated by considering the volumes of records to be handled in transaction files, master files, and intermediate files, with reference to the read/write time, for the tape deck system concerned, and the block size employed for files handled in each run.

Reference should again be made at this point to Fig. 6.4; the timing chart shown in this diagram can be used to determine input/output times provided block size and overall file capacity in words are known. The rewind times are also shown.

Example – Paper Tape Input

Assume that records punched into paper tape have to be read as part of a daily order processing system. Assume further that 5,000 such orders arise daily and each may consist of up to fifty items punched in the following format:

DESIGNING FILE SYSTEMS—MAGNETIC TAPE

		Characters
Customer No.		5
Customer order No.		6
Date		6
Item No.	repeated up to a maximum of 50 times; average 16 per order	3
Quantity		3

Assuming that the average number of items ordered per transaction is 16, on average each transaction will require a total of 113 characters. Overall, the daily volume will be 560,000 characters, and with a tape reader operating at 1,000 characters per second the theoretical input time will be 9 minutes approximately. In practice, paper tape is often prepared in lengths which provide a convenient handling volume for punching and verifying, and therefore it is necessary to include an allowance for the input and respooling of the individual tape lengths. On the basis that each batch will contain fifty orders, the overall peripheral time for the tape reader must include, say 100×1 minute, for handling the batches.

Example – Magnetic Tape File Handling

Assume that a personnel records file has the following information for a company employing 10,000 salaried staff:

	Format	Characters	Words
Staff no.	BCD	6	9
Name	BCD	30	
Address	BCD	32	8
Grade	BCD	4	1
Leave scale	BCD	1	2
Sex	BCD	1	
Date of birth	BCD	6	
Qualifications	BCD	20	5
Department	BCD	4	1
Salary	Binary		1
4 × previous salary charges	Binary		4
Current job description	BCD	16	4
4 × previous job descriptions	BCD	64	16
Internal Telephone	BCD	4	1
Home telephone	BCD	8	2
			54

The record size we can assume in this case is fixed at 54 words, with 10 records to a block making the block size 540 words.

The overall size of the file is 54 × 10,000 = 540,000 words, and by reference to the chart in Fig. 6.4 we can see that the overall time to read or write the file is something under one minute.

Elapsed Peripheral Time per Run

Having determined the individual peripheral times for a run, the analyst has to contemplate the possibilities of running all the devices concurrently. This may depend upon the hardware input/output channels available or upon the relationships of data conditions arising in the run. The hardware restrictions will be known for a particular configuration, and can be established by reference to manuals provided by the computer manufacturer.

Data restrictions particularly arise where slower peripherals are hooked up to magnetic tape processing runs; here, bursts of activity requiring, say, printed or punched output may arise in such a manner that magnetic tape operations are momentarily delayed. This kind of activity may be related to the hit rate of transactions to corresponding main file records: the analyst can estimate the frequency of these occurrences and the volume of activity resulting from this condition.

If runs are well designed and if input/output buffers are used to balance peripheral and processing activities as previously described, it should be possible to arrange that the overall elapsed time is equal to the peripheral time required by the longest running peripheral.

Processing and Overall Running Time

Processing time includes the time required to perform the desired logical operations on data in the run plus the time required by the supervisory system to control peripheral transfers and to accept and deal with data transferred to and from the input/output units.

The actual data processing time has to be estimated by establishing the number of operations performed in each routine, and thus an average processing time per routine is ascertained. Then estimates must be made of the average number of times that particular routines are activated in the run.

The executive or supervisory times will be related to the number of transfer operations necessary to get data into and out of the processor in the run. If the number of transfers (e.g. blocks) is known, this factor can be multiplied by constants derived from the manufacturer's supporting technical literature to ascertain overall processing time attributable to these operations.

For a fully time-sharing system, the overall running time for a run can be determined by comparing the elapsed peripheral time with processing time: the overall time for the job will be equal to the greater of these two factors, i.e. the run is either peripheral-limited or processor-limited. Of course, when a program is being run along with others in a multi-programming environment, the overall time depends upon the relative priorities and characteristics of the other programs; and the time spent in switching from one job to another becomes an additional overhead.

Chapter 7

Designing File Systems— Direct Access

THE PRINCIPLES OF DIRECT ACCESS STORAGE

In the previous chapter the basic principles of file handling when using magnetic tape were discussed. It has been shown that magnetic tape is a serial storage medium, and that, as a consequence, it is necessary to batch transaction records before updating a file. Thus information derived from a magnetic tape file is relevant only when qualified by considering the time at which the file was last updated.

Direct access storage media has a number of advantages over magnetic tape systems, but mainly it is distinguished by the ability to update records or access them for on-line inquiry without searching one by one through the records in the file concerned. The time needed to extract a particular record is measured in milliseconds, and the access time is not critically dependent upon the location of the record previously addressed.

Thus the term *direct access* is derived from the nature of the systems that are potentially available with such storage devices. Chapter 8 of this book introduces some examples of this type of system.

Direct access devices are not used solely for on-line systems; they may also be used for conventional batch processing applications in which advantage can be taken of their characteristics, e.g. to speed up sorting, to facilitate more efficient indexing operations, or to obviate the need for sorting main files entirely.

Direct Access Devices

Direct access devices, sometimes referred to as random access devices, include magnetic drums, magnetic card files, exchangeable disc stores, and fixed disc stores. Each of these backing stores are intended to supplement the main memory capacity of a computer by providing fast access to transaction files, master files, history files, output files, program libraries, indexes and tables,

DESIGNING FILE SYSTEMS—DIRECT ACCESS

and executive or supervisory software. Although the mechanics may differ from one device to another, the principles are the same. In a magnetic drum, the drum is revolved continuously and data stored upon tracks around the circumference of the drum are available at fixed points in the basic cycle time. If each track has a single read/write head, the average access time for any unit of data is half the time required to complete one revolution. A magnetic drum might hold more than 200,000 words, and it is possible to connect several drums to a central processor. Operating at about 7,000 revolutions per minute, a high speed drum might provide an average access time of a few milliseconds and have a transfer rate as high as 10 million bits per second.

A typical disc store consists of a number of rotating discs each coated upon both surfaces with a magnetizable material. Information is written to, or read from, the disc by a series of read/write heads – one for each surface – which can be positioned above certain areas of the corresponding discs according to instructions received from the program. Each disc surface contains a number of recording tracks which are further subdivided into sectors or bands. The instructions from the central processor can cause the read/write heads to be positioned to read or write from specified tracks. As with magnetic tape, information is generally recorded as a series of 6-bit characters each with an associated parity bit. A small device of this type would hold about 8 million 6-bit characters and a large device might hold 400 million characters; access time to any chosen bit might be between 12–150 milliseconds.

A typical magnetic card file consists of, say, 20,000 or more individual cards which are coated with a magnetizable substance. The cards are stored in magazines from which they may be extracted individually, according to instructions given by a program, and transported at high speed past a read/write head. Each card might contain about 150,000 individual characters arranged into, say, sixty bands. The access time to individual cards varies, perhaps, between 300 and 450 milliseconds per card, but usually once a card has been accessed it can be circulated continuously past the read/write head until directed by the program to return to its original position in the magazine. Whilst the card is circulating, data can be transferred from it, or to it, at speeds of the order of 80,000 characters per second.

The detailed techniques for handling files on direct access media are, of course, related to the particular device used. Throughout the rest of this chapter, the examples chosen will be based upon the use of a disc store; the same basic techniques apply to the other devices with minor variations.

Basic Modes of Processing

There are three modes for file processing when using direct access stores:

    (a) Serial processing;

(b) Selective sequential processing;
(c) Random processing.

Serial processing implies that a file is processed by reading from the first physical location of the file to the last location for that file, taking each physical block in turn. As we have seen with magnetic tape, this technique implies that transactions are batched and sorted to sequence before being applied to their relevant master file, and a new master file is created on each updating run. It would be possible to update direct-access files in this manner, but generally it is not fruitful to do so, and serial processing is reserved for special situations, e.g. when it is necessary to edit every record on a particular file.

Selective sequential processing takes place when a transaction file, sorted to sequence, is applied to a master file. The program selects master records according to the keys of records appearing in the transaction file. Master file records are examined if a corresponding transaction arises only, and the record retrieved at any particular time will not be necessarily physically alongside the record previously addressed, although the master file records will be selected in sequence.

Random processing implies that transactions may arise in any sequence, and access is therefore required to any record on the master file without regard to the record previously addressed.

Basic File Storage Concepts

First, let us consider further the way in which a disc store is used. Each file is not considered primarily as a number of data blocks, but instead is considered as one, or more, seek areas. A seek area is an area of storage which can be searched without physical movement of the read/write head across the surface of the discs.

An exchangeable disc might consist of six separate magnetic discs with the read/write heads arranged as shown in Fig. 7.1. Assume that all the heads must be moved in unison. Once the heads are positioned, access can be achieved to data on each surface of each disc without needing to reposition the heads. The area thus available includes bands on each disc surface, and blocks of information can be selected from particular bands by switching between heads. The area thus available is known as a seek area; access time to a particular record within a seek area is, on average, equal to half the time needed to complete one revolution of the disc, say in the order of 12 milliseconds.

If the read/write heads are repositioned to cover other bands, then a further seek area is available, and so on. The system designer has to organize his files in such a way that they utilize this concept to minimize the amount of head

DESIGNING FILE SYSTEMS—DIRECT ACCESS

movement necessary in handling files. For an exchangeable disc store, the average access time applicable between records on different seek areas might be in the order of 75 milliseconds. This method of working applies, in one form or another, to all the direct access devices previously mentioned, e.g. in a magnetic card file the seek area is one card.

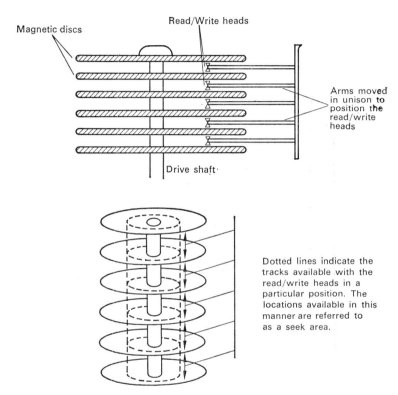

FIG. 7.1. Outline of a magnetic disc store.

A seek area is subdivided into units referred to as buckets. Sometimes a fixed bucket size applies, but usually the size can be varied provided a constant size is used for a particular file. A bucket contains one or more blocks of information, up to a maximum perhaps of 16 blocks.

The bucket is the basic unit of transfer between a direct access store and the central processor, and can be used to contain one, or more, file records according to the size of the records and the size chosen for the bucket itself. The maximum bucket size is determined by the characteristics of the particular device and maybe by the associated software.

Thus a direct access store can be considered as a number of seek areas of predetermined size which may be assigned to particular files, and within each file area records are organized as required into buckets.

Within a file, each bucket is given a logical number; these are applied in ascending sequence from one seek area to the next for all buckets of the file.

File Mapping

Because the records of a particular file may be spread over a number of disc surfaces, a special technique has to be adopted for recording the extent and

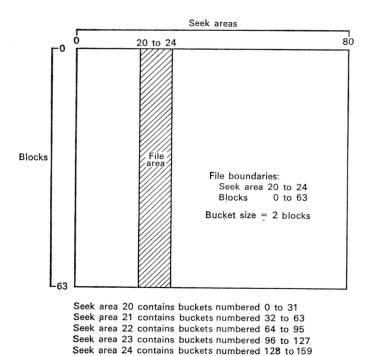

Seek area 20 contains buckets numbered 0 to 31
Seek area 21 contains buckets numbered 32 to 63
Seek area 22 contains buckets numbered 64 to 95
Seek area 23 contains buckets numbered 96 to 127
Seek area 24 contains buckets numbered 128 to 159

FIG. 7.2. Example of a file map.

organization of files. The technique requires that a map is drawn on the lines of that shown in Fig. 7.2.

Seek areas are numbered and shown subdivided into blocks which are the basic units of transfer between the device and the central processor. The relationship between blocks and buckets has to be specified for the file concerned, and the seek area and blocks allocated to the file can then be given. For example, in Fig. 7.2 the file is shown as occupying 5 seek areas, all blocks

DESIGNING FILE SYSTEMS—DIRECT ACCESS

in these seek areas are utilized, and 2 blocks are assigned to each bucket. Thus, per seek area, there are 32 buckets, making a total of 160 buckets for the file concerned. The buckets are numbered from 0 through to 159 as shown, and these numbers are used as addresses for the file concerned.

This is a simple example; in practice much larger files arise, and a file may need to be assigned to seek areas that are not consecutively numbered, and some seek areas may need to hold parts of different files.

It is probably worth elaborating the relationship between blocks and buckets. A bucket is a programming concept and its size is related to the application concerned but is constrained by the software standards that apply; e.g. maximum bucket size may be 8 blocks. A block is a hardware concept, and is usually fixed for a particular storage device; e.g. 1,024 characters. There is not much point in setting bucket size lower than block size since a block is effectively the smallest amount of data that can be retrieved.

File Packing Density

A file consists of a number of buckets each of which may contain one or more records. It is possible that the records may be packed into every bucket so

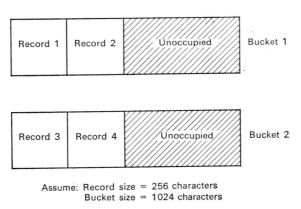

Assume: Record size = 256 characters
Bucket size = 1024 characters

FIG. 7.3. Buckets loaded at packing density of 50 per cent.

that every possible character position is utilized, and the file packing density would therefore be 100 per cent. However, no matter whether fixed or variable-length records are used, it is unlikely that such a condition could exist with a main file – usually one has to leave space in buckets to receive new records or expand the existing ones.

When one creates a file it is usual to estimate its potential expansion and to initially load records into buckets in order to accommodate this growth. In the example shown in Fig. 7.3 the records are loaded in such a manner that

SYSTEMS ANALYSIS IN BUSINESS

50 per cent of each bucket remains unoccupied. At some point in the future, certain buckets will overflow, and records that would otherwise be loaded into them are then consigned to a special overflow area. Eventually the buckets in the overflow area may become full, and the file has to be reorganized so that a larger storage area is allocated to the file; once more file growth is anticipated in setting the packing density.

This aspect is one of the major considerations in planning files on direct access equipment, and we shall return to it in more detail.

TYPES OF FILES

Serial Files

A serial file is one where the records are stored in successive locations within each bucket and every bucket in the file area is filled. The records may or

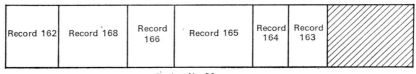
Bucket No 90

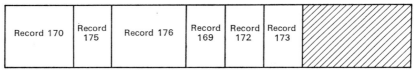
Bucket No 91

FIG. 7.4. Arrangement of records with buckets of a sequential file. (Note: The bucket index would be arranged to show records 161 to 168 in bucket 90, and 169 to 176 in bucket 91.)

may not be sorted into sequence. Transaction files are often held in this manner, and output files also may be held in serial form. Access to records is made bucket by bucket in the sequence of the bucket numbers allocated for the file. The bucket packing density in a serial file may be 100 per cent.

Sequential Files

In sequential files, records are assigned to buckets in groups, and generally the group in a particular bucket will have keys higher than the group in a lower numbered bucket, but within buckets the records may not be in key sequence. The highest significant key in each bucket is used as a pointer to the records held within it, and an index is established to directly select the bucket relevant to a particular transaction (Fig. 7.4). To address a record stored in a sequential file, it is necessary first to establish the keys sought, then derive by means of the index the bucket concerned, read the bucket into main memory, and search its contents for the desired record. If the record is updated the bucket must be written back to the direct access store once more.

When updating a sequential file, transactions are selected one by one from a serial transaction file and are applied to the main file selective sequentially. Buckets relevant to the current transaction records only are accessed, and provided the transaction file is sorted into sequence it is not necessary to access a bucket more than once. Sequential files can also be randomly processed, unsorted transaction records being applied as and when they arise to select buckets via the index. This method will entail selecting records and buckets up and down the file from one seek area to another. A sequential file will usually be updated in random manner if the transaction hit rate per seek area is very low, and if a relatively fast response time is entailed.

Random Files

In random files records are assigned to buckets by some logical process which can be repeated to retrieve the buckets concerned and select the records required on demand.

The method usually entails the manipulation of the keys of each record in order to create a bucket address, and thereafter for a given key the bucket required can be located and searched to find the corresponding main file record. This technique of creating bucket addresses from keys is known as *address generation*.

It can be expected that some difficulty will be encountered in developing an address generation technique relevant to a particular set of keys. The systems analyst rarely has the opportunity to design the record keys that his system will operate upon because these matters are often determined by the previous history of the organization, its products, and procedures. Ideally, the random processing technique should generate unique addresses for each record, and yet yield a relatively high file packing density, but in practice a far less efficient condition must be accepted.

Address generation methods usually assign a high number of records to some buckets, and a low number to others. For this reason overflow may arise frequently although the overall file packing density may be relatively low.

A random file can be processed randomly allowing single transaction records, or small batches, to be used to inquire and update the file directly as they arise. Such techniques often permit a faster response than random processing against a sequential file, because it is not necessary to first apply the transactions to an index. Random files are often used in real time situations, where the volume of traffic is comparatively low but where a quick response is needed.

SEQUENTIAL FILES AND INDEXING

The method most popularly used for accessing via an index is known as *partial indexing*. It is suitable for large files and those in which the key values are unevenly distributed over the possible range. The index tables are often maintained on the direct access device, to reference the keys of transaction records to the buckets where the relevant main records are stored. To update a main file record the transaction key is isolated, then a search is made through the index to find the key value and its cross-reference to the bucket address of the main record. The relevant bucket is then read into main memory, searched, and the required record is selected and processed as necessary, and the whole bucket written back to the direct access device once more. Sometimes there are several records stored in a bucket, and if the records are allocated to buckets sequentially, as shown previously in Fig. 7.4, the index will show the key of the highest record in each bucket only.

If a file occupies several seek areas each containing several thousand records, a method is needed to organize the index in order to minimize the retrieval time. It is quite usual to have a seek area index which indicates the highest and lowest key value in each seek area. Each seek area then might have a bucket index indicating the highest key in each bucket of the area.

A seek area index, if small, could be retained in main memory and perhaps the bucket indexes could also. However, if the file is large the bucket indexes will probably be called into main memory as required when transactions relating to a particular seek area are processed.

RANDOM FILES AND ADDRESS GENERATION

Random files are employed where transactions arise singly and in such a manner that any of the main file records may need to be addressed regardless of the record last processed. The retrieval method depends upon generating a bucket address by manipulation of the key of the transaction record to produce the address of a bucket containing the main file record.

Address generation techniques should ideally generate a unique address for each record by manipulation of the key. The process used must be such that it can be repeated for any key in the series. The method must also provide an economic distribution of the records over the file area, with the packing density approaching 100 per cent. These ideals are rarely met in practice;

DESIGNING FILE SYSTEMS–DIRECT ACCESS

usually some buckets are under-utilized while others are assigned more records than they can store.

There is a great variety of address generation methods and usually what suits one application will not do for another; the following steps are often involved:

(1) The record keys are processed to produce results evenly distributed over the set of numbers;
(2) The results obtained are expanded or edited logically to correspond with a range of bucket addresses available;
(3) A base number is added or subtracted to derive actual bucket numbers.

The systems analyst is rarely able to specify his own key numbers; usually the keys are determined by the previous history and nature of the organization.

THE PROBLEM OF BUCKET OVERFLOW

We have seen that when a file is initially created, it is necessary to allow for its future expansion; sufficient storage is allocated to permit records to be expanded or new records to be added. The analyst must determine how much storage can be economically allocated to a particular file: on the one hand, he should try to ensure that storage capacity is not wasted, and, on the other, that the file is sufficiently well organized to minimize the time needed to retrieve records.

In theory, the file could be organized to permit every conceivable record within the range of keys concerned to have a unique storage location. However, one often finds that only a small percentage of the possible keys in a particular range are utilized. For example, an inventory might consist of 30,000 line items each of which have been allocated an identifying number in the range 2–3 million; the numbers currently allocated may be unevenly distributed throughout this range, and perhaps new items can be expected to arise randomly in future.

If the analyst knows that between 30,000 and 36,000 records are likely to suffice for the next year ahead, he will need to allocate an appropriate number of buckets to contain the file. At four records per bucket, a capacity of 9,000 buckets will be necessary – much less than that needed to encompass the whole key range.

Since it is hardly ever possible to predict the actual key values that will arise, there is a high probability that the need will arise to pack more than four records into particular buckets while others remain underpopulated. This situation could occur whether the file was randomly organized or sequentially organized.

SYSTEMS ANALYSIS IN BUSINESS

Overflow in Sequential Files

In practice, a sequential file is usually created in such a way that the records are initially distributed evenly throughout the buckets assigned to the file; each bucket being filled up to a level specified and having a certain percentage of capacity free for future file expansion. Some of the buckets assigned are specified as overflow buckets and are kept empty when the file is first created. Overflow occurs subsequently when particular areas of the file become conjested due to additional records being added or existing ones expanded.

When overflow occurs in a bucket it is necessary to store one or more of the records assigned to it in another bucket, usually set aside as part of an overflow area on the same seek area. Thus when retrieving overflow records it is not necessary to incur a longer access time for movement of the read/write head.

The problems of file organization and creation are usually handled using the computer manufacturer's software, but the analyst will need to understand and use this software intelligently.

Overflow in Random Files

With random files, overflow is much the same as in sequential files except that it can occur when the records are initially loaded into the file area. With sequential files, special buckets are assigned to receive overflow records on each seek area; but this is not sensible for random files. Address generation techniques often result in a very uneven distribution of records, and many home buckets are likely to remain unused due to these limitations in the address generation method. It is more efficient for overflow records to be assigned to these unused home buckets, if this is possible. Thus home buckets are also loaded with overflow records, which may cause more potential for overflow. Overflow depends upon the distribution of the keys to be handled, but the effects of overflow can always be mitigated by storing overflow records on the same seek area as their home buckets.

Retrieving Overflow Records

When a record is placed into an overflow bucket a stub (or tag) record is left in the original home bucket quoting just the keys of the record and the address of the overflow bucket (Fig. 7.5). This process is known as tagging and is commonly used.

A program is thus able to retrieve the overflow bucket and search for the record concerned. It is possible that eventually so many stubs are placed into a home bucket that one, or more, of the original records have to be removed and stored in the overflow area in order to accommodate the stubs. In an extreme case a home bucket might contain overflow stubs only, but usually this condition is detected by special programs supplied by the computer manufacturers to analyse and reorganize direct access files.

Overflow records could eventually occupy the whole overflow area. Software systems to safeguard users against loss of data are provided; this will store records on special reserved areas when a file is suffering from such severe symptoms. Processing a file in this condition takes more time since many additional seek times are incurred.

The systems analyst working with a particular software system for controlling direct access files will decide the size of file areas to be allowed and will seek to minimize the processing time caused by overflow, and yet achieve efficient use of storage areas on the device.

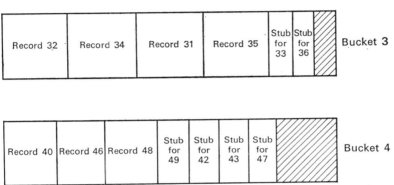

FIG. 7.5. Overflow stubs in home buckets.

File Reorganization

File reorganization systems are provided by the computer manufacturer. A sequential file can be reorganized by copying to another file area which is larger than the original. In so doing, the overflow tags are read and the corresponding records are retrieved and placed in home buckets once more. The file reorganization is carried out to a specified bucket packing density and a new index table is created for the file.

The analyst must specify the storage capacity necessary when setting up a file or reorganizing it. He must bear in mind the growth rate both in the number and size of records stored so that he can allocate an appropriate overflow area.

The difficulties in making estimates of this nature, rest upon the initial problems in assessing the rate of file expansion. Observation of existing systems can produce reasonable estimates of the number and frequency of new records arising and for deletions of redundant records.

Activity may be highest in certain parts of the file, and it may be possible to extend some file areas at intervals in a sequential manner. Also some seek areas can be assigned a lower packing density than others.

BASIC TIMING PROBLEMS FOR DIRECT ACCESS FILES

So far we have discussed the principle files and modes of processing that are available for direct access devices. The choice of a particular method has to be made by the systems analyst considering the following factors:

(a) The response required by the users of the system;
(b) The costs of file maintenance and processing in terms of computer time;
(c) The utilization of storage capacity;
(d) The complexity of the programming work needed to implement the system.

The response required is of overiding importance, and once this has been established the analyst has to investigate various ways of achieving this objective and try to optimize the other factors. In the long run the costs of operating the system on a day-to-day basis will probably assume more importance, although storage utilization is ever an important consideration since the capital cost of direct-access backing stores is high. The evaluation of a particular system inevitably entails the reconciliation of these conflicting constraints, e.g. if the packing density of a file is low, poor storage utilization results; but the percentage of overflow records is diminished, and thus less time is needed in retrieving records and writing them back to the storage device.

TIMING METHODS FOR FILE PROCESSING

Let us examine this argument more closely by considering the three basic modes for processing sequential files.

Serial Processing

When a sequential file is processed *serially*, every bucket must be read, and if an amendment is made to any record within a bucket the bucket must be written back to the storage device once more to preserve its contents. Thus every seek area is accessed, every bucket is read (including overflow buckets), and every bucket containing a record that is amended must be written away once more. The basic timing formula for processing a file serially is therefore

$$(Sn \times St) + (Bn \times Rt) + (Br \times Wt),$$

where Sn = number of seek areas, St = seek time (minimum), Bn = number of buckets, Rt = bucket read time, Br = number of buckets containing hit records, and Wt = bucket write time.

Selective Sequential Processing

Here the pattern is a little more complex since main file records are only retrieved where there is a transaction arising in the period under review. Only some seek areas may need to be accessed, and only some buckets within these seek areas are read and written back to the storage device. For each transaction it will be necessary to retrieve an index bucket in order to look up the bucket address of the desired main file record, but this time can be minimized if the index for each seek area is stored in turn in main memory.

Overflow conditions must also be allowed for; as a minimum it will be necessary to read and write to the overflow area each time a tag is encountered in a home bucket. If new overflow conditions arise during the run it will be necessary to place records into an overflow area of main memory which must be written away to the next available overflow bucket. The overall time for processing a file selective sequentially is, therefore, $(SR \times St) + Br(Rt + Wt) + (SR \times RIt) + OB(Rt + Wt) + NOB(Wt)$, where SR = number of seek areas hit, St = seek time (average), Br = number of buckets containing hit records, Rt = bucket read time, Wt = bucket write time, RIt = read time for retrieving index buckets, OB = number of tags encountered in home buckets, and NOB = number of new overflow buckets written in the run.

Random Processing

To process a file randomly it is probably best to assume that each transaction will arise independently of any other, and thus will probably imply the selection of a seek area on each occasion. Having accessed the seek area the relevant data bucket must be selected for updating and then be written back to the storage device once more. Therefore, the basic timing formula for the whole file is

$$(T \times St) + T(Rt + Wt),$$

where T = number of transactions, St = average seek time, Rt = bucket read time, and Wt = bucket write time.

If an index is used to select main file records rather than an address generation technique, then further allowance must be made for reading the bucket index from each seek area $(T \times Rt)$. An estimate must also be made for the percentage of overflow tags that will be hit and an additional read and write operation allowed for in each case.

SYSTEMS ANALYSIS IN BUSINESS

The Relevance of Timing Estimates

The timing calculations suggested above are very rudimentary and serve merely to indicate the basic factors to be considered when evaluating design proposals. To a large extent the timing will be affected by the manner in which the computer manufacturer has designed and implemented the general purpose file handling software used for the storage device. For example, when handling a sequential file if only one input/output buffer is provided, the current home bucket will have to be written back to the storage device each time an overflow bucket is retrieved. After handling the overflow condition, the home bucket may then be read once more to continue processing other transactions that may relate to it.

Also, it must be confessed that some factors are difficult to assess, e.g. the number of buckets hit when updating a sequential file selectively. The times attributable to overflow processing are similarly complicated – it is difficult to estimate the number of tags that will be hit in an updating run or the number of new overflow situations created.

The literature provided by computer manufacturers can be helpful, and often statistical tables and graphs are presented which enable sufficiently accurate estimates to be made.

The nature of the application itself determines the approach in most cases. For example, random processing is reserved principally for on-line applications where the access time is all important. In such situations it is likely that some or all of the computer resources will be dedicated to the application and the storage capacity can perhaps be utilized without over emphasis on the file packing density. The nature of the keys and their distribution over the range of values concerned will influence the choice between address generation and indexing as methods of access when processing a file randomly.

In batch processing operations the different functions involved will dictate the type of processing. Where all records of a file have to be examined (e.g. editing, validating, selecting operations), a file will probably be processed serially. To update a file in batch processing mode it is best to use selective sequential processing, unless the hit rate for main file records is very high, in which case serial processing might apply.

Chapter 8

On-line and Real Time Systems

THE NATURE OF ON-LINE APPLICATIONS

Most of the examples developed so far in this book have related to batch processing applications in which data is collected and prepared in accordance with some predetermined schedule before being applied to the computer. The user generally receives output according to some agreed specification, but has the ability to also specify *ad hoc* outputs. He may also receive information relating to exception conditions specified perhaps as parameters whenever the reporting subsystem is operated. In describing direct access storage equipment it was shown that one of the prime advantages arising from this type of backing storage is the ability to retrieve data quickly without searching through all the records in a particular file. Thus transaction records can be handled as they arise, and the need to batch transactions and to sort them before processing is obviated.

It therefore follows that systems can be designed to:

(a) Make random inquiries of files and present selected data to users as and when required;
(b) To make amendments to data within records in direct response to transactions as and when they arise.

Thus computer users are given the opportunity to have a more dynamic relationship with data stored in the file system. To take advantage of these characteristics, the users also need data terminals equipped with hardware to input transactions and to display results. Some software will also be needed to control communication between each terminal and the central computing system, and if the terminals are stationed at distant locations, data transmission equipment is necessary.

The cost of this additional hardware and software has to be evaluated and the benefits of applications using these facilities must be considered in relation to the cost incurred.

The hardware available for remote terminals includes teletypewriters, visual display units, graph plotters, as well as more conventional peripheral devices such as card readers and punches, paper tape readers and punches, and line printers. Teletypewriters and visual display units are particularly well suited to interactive systems since not only do they provide facilities to display data, but they are also equipped with keyboards to enable transactions to be entered into the system.

Punched-card and paper tape peripherals, on the other hand, are intrinsically designed for operations in batch processing mode. Where they are used as remote devices it is generally to extend the facilities of a central computing system enabling batches of data to be input or output at remote locations.

It is unlikely that an organization will desire to run all its applications in on-line mode; it is common to have one (or more) basic applications operating in on-line mode while other work is run in batch processing mode.

On-line applications are usually implemented to provide users with a fast response to their inquiries. The system may also be specifically intended to capture details of transactions as they arise, so that certain files contain an up-to-the-minute picture of events.

The application itself dictates the pattern of the system and the nature of its response to individual users. On-line systems can be categorized as ranging from truly real time systems, which are extremely time critical, to remote inquiry systems which may be able to tolerate delays of minutes before giving a response to the user.

Perhaps a more helpful method of categorizing an on-line system is to consider the relevance of the data that it handles. For example, a management information system may require that certain managers have access to data showing current conditions. The users here are not normally interested in particular transactions or events, but in seeing an overall picture or trend for the year to date or in a particular month. The data entering the on-line files may itself be derived from batch processing systems which deliver summaries to the on-line data bank at daily, weekly, or monthly intervals. The response to subscribers demands may need to be speedy, but file updating and transaction handling are not critical activities.

Real time systems, however, are centred around transaction based applications; the most well-known examples are air-line seat reservation systems in which hundreds of subscribers may be attempting to book seats at the same time. The individual seats or resources can be considered as items in an inventory which must not be allocated to more than one user. Other batch processing applications may also be operated as background to a real time system, and usually will be initiated automatically whenever the real time system is operating at a low level of loading.

ON-LINE AND REAL TIME SYSTEMS

TIME SHARING

In order to handle a large number of terminals simultaneously, as well as conventional peripherals in batch processing operations, the computer has to be equipped with special hardware and software facilities. The term time sharing is used to describe these overall characteristics, but the terms multi-programming or multi-processing are also used.

Time sharing has many meanings and is one of the most overworked terms used in data processing. It is used at its lowest level to describe situations where a program is able to drive two or more peripherals simultaneously, and also to refer to the situation where a processor and its executive program can control several independent programs concurrently. The term also has a wider meaning, implying that a central computer system controls a number of terminals in a data communications network; each user has the ability to call up the central computer to occupy some of its resources for a while according to particular requirements.

Both batch processing and on-line systems may employ time-sharing techniques, but the distinction between them is usually related to the response time between the system receiving transactions and providing results. An on-line system is so designed to provide one or several users with high speed information service, or to also provide them with computing power when needed without significant delay. An on-line system may therefore be considered a highly developed application of the techniques of time sharing providing for close interaction between man and machine.

MULTI-ACCESS COMPUTING

Computers are capable of storing vast quantities of data, but a man can only digest a small amount at any one time. Thus there is an argument for providing numerous channels for communicating with the computer so that its resources are efficiently utilized. Several users may have access to a particular terminal, and perhaps a hundred or more terminals may be served by the central computing system. This sort of general purpose facility is characteristic of that needed in scientific or research establishments where individuals use the computer to tackle a variety of tasks.

The term multi-access has been used to describe this situation. Several users need access to computer files and processing power at various times and for purposes related to their particular function. For example, at any instant some users will need to run programs to update files, others to process data in files, others may be compiling programs or performing numerical computations upon data. Each user may have a choice of several languages according to the nature of his application, and he may have private secure data files stored within the system as well as access to other files according to security standards inherent in the system.

A multi-access computer system is not dedicated to one basic function as is a reservation system, but, instead, it controls many jobs both in on-line and batch mode. The response requirements are critical, and the executive or supervisory program has to be developed as a general purpose software system able to handle the requirements of all users. Large software items of this nature are usually developed by the computer manufacturer.

Such systems have to provide a variety of languages and functions according to the needs of different users. Installations of this type are found in research institutes or universities, where jobs are of a scientific nature requiring complex processing but perhaps little file handling. The average time for jobs operated in this environment is usually a few minutes. However, multi-access computers are also found in business houses and manufacturing or distribution centres, perhaps catering simultaneously for the needs of accountants, managers, scientists, designers, engineers, economists, and technologists of one kind or another.

Some multi-access computers are partially dedicated to particular applications, perhaps process control, while other tasks are performed as background work being allowed to occupy the processor whenever the main job is quiescent. Very large time-sharing systems can be designed to perform real time and other tasks, and the future will undoubtedly be in this direction with multi-access computing becoming a standard requirement for all, except the smallest, machines. In the future, increasing dependence upon hardware and software of this type may imply two or more processing units each capable of driving any of the terminals and their input/output devices. The need for precautions of this nature is not so much dependent upon the need to safeguard particular applications, but to ensure that a continuing service is maintained to the community as a whole, many individual functions and services being dependent upon the maintenance of the computing system.

RELIABILITY IN ON-LINE SYSTEMS

The use of on-line systems implies greater interaction between men and machines, and also therefore greater dependence upon the computer hardware and software. We have already defined real time systems as being situations in which the failure of the system is potentially disastrous for the application concerned. Therefore it is clear that exceptional precautions have to be taken to avoid failures or to minimize their effects.

Failures can be divided into two major categories – hardware and software. Hardware failures can be covered by having back-up facilities, e.g. two or three independent power supply sources, two processors, several peripherals of each type, and so on, but software faults are often difficult to minimize since they may arise only in certain circumstances and be attributable to rare data conditions. In major software items, tracing and correcting errors can be a long and costly task. Errors in executive or supervisory software may severely cripple a computing facility, either putting it out of action completely or severely reducing its powers.

All on-line systems have a requirement for reliability, and this dependence for reliability increases as we approach the truly real time application. In multi-access computing installations the dependence is perhaps not quite so critical, but nevertheless assumes an increasing importance where a large number of users are involved. The users of multi-access terminals have the power to correct and test their own programs on-line using the terminal without impairing facilities available to other users; this may provide some reassurance, but it should be noted that each user is entirely at the mercy of the trusted programs that exercise control over individual user programs and files.

REAL TIME APPLICATIONS

The essential nature of a real time system is one of speed in which the response required is directly governed by the application itself. It should also be observed that the survival of the system is essential to the whole operation that it controls, and system failures potentially have a devastating effect on the operation.

A real time system is one dedicated to a single application, as in the airline seat reservation system; many terminals must be connected to the system to enable several users to potentially have access to the same data at one time. The transactions that arise are of a more or less standard nature, and the whole system is primarily geared to repeatedly carry out the same processing function.

The basic processing problem may be relatively simple, but usually the application itself is a vital part of the organization's operation, and system failures are intolerable. This usually implies a very quick response from the system and a very complex executive software system to protect data files and individual transactions from equipment failure.

In real time systems the user is typically one of the many who may be concurrently trying to reserve resources, and it is undesirable that any particular resource should be allocated to more than one user, and delays must be minimal if the system is to be viable. Reservation systems of this type have to respond within 10–15 seconds to any subscriber, but in some real time applications it might be feasible to queue transactions so that, on average, access to resources is given within, say, 1 minute. In some inventory control applications, for example, such a delay might be acceptable, but if the average delay is too long, queues of transactions will begin to form at the terminals.

In other applications the response time may be exceedingly critical: air defence systems are an example; here unidentified aircraft have to be intercepted before they can approach the defence area, and valuable seconds are needed to get interceptors directed to meet the intruder. This degree of reaction is provided by a machine-to-machine interface in which the central computer is connected direct via transmission lines to a number of remote radar detectors.

Software reliability can only be attained by foresight and diligence in developing and testing the product. In the case of a multi-access executive system the computer manufacturer will be responsible for providing the software, but probably a number of major users will co-operate in the testing and debugging of the initial facilities. The final product will be a general purpose trusted program which can be implemented progressively.

FAIL SAFE

In real time applications, the task is far more specialized; the executive program is essentially adapted to the requirements of the one major application. Not only must back-up facilities be available, but special methods of working have to be observed when a breakdown arises. One can almost regard a real time system as being one large and specialized executive program which incorporates a variety of exception routines to safeguard the operation of a relatively simple transaction processing system.

Every transaction is vital, and the recovery procedures have to be developed to enable each outstanding transaction to be off-loaded to a safe storage area. All files and data have to be protected in this way during the phase in which the breakdown occurs. After a breakdown the next phase is an immediate reconstruction of the conditions existing at the breakdown before a restart is initiated. Control of this nature is attained by skilful systems engineering; the whole of this process may have to be conducted within a matter of a few seconds in which the operator may be taking necessary action to re-establish the system by switching hardware resources to permit continuance of the application. While a breakdown is in process, action has to be taken to queue transactions that parts of the system may be still receiving. The restart time is critical since the system may then be loaded instantly to its peak level, with the queued transactions, while operating in crippled mode. It can be seen, therefore, that the application design has to be skilfully conducted, and the investment in hardware and skilled staff may be very high if the application is to be successful.

Chapter 9

Control of Systems Development

STAGES OF SYSTEMS WORK

In the first chapter it was shown that although different organizations may approach the work of systems analysis from different angles, nevertheless it can be stated that the following activities do take place in the development of most applications:

(1) Project identification and selection;
(2) Investigation of existing procedures;
(3) Feasibility study;
(4) Design of files, programs, and data collection procedures;
(5) Documentation and presentation to management and staff education;
(6) Program writing and testing;
(7) File creation;
(8) System testing;
(9) System operation and monitoring.

In extreme cases a single analyst might be expected to conduct each step of this program, but in most situations there will be a separation of the functions to obtain the best results from the skill of individuals. The most obvious example of this division occurs where a separate programming section deals with step (6), and may also help in steps (7) and (8).

In any project, except the very simplest, it is usual to find a great variety of interests involved. The systems department has to coordinate these interests and guide the project to a successful implementation, as well as supply its own particular skills in designing files and computer procedures. In preceding chapters we have discussed some of the ways in which a systems analyst conducts his job, stressing mainly the techniques and principles that he must observe in going about a task. This chapter concentrates mainly upon the way in which projects are controlled within the organization as a whole, laying particular emphasis on the important task of selecting projects and ensuring that they are pursued in a manner likely to be beneficial to the achievement of the business objectives of an organization. Some remarks about long-term planning for the systems department are then given, a discussion of documentation standards, and, finally, some notes about control of the programming functions.

GETTING THE OBJECTIVES RIGHT

One of the most vital aspects of systems work lies in getting the basic objectives of an assignment established and agreed by all concerned. This stage of the analyst's work depends upon a real appreciation of the line management's problems, and if an understanding can be made at this point much has been done to create confidence throughout the subsequent stages of the project. It is also important for the systems department to explain to their user departments the manner in which an assignment will be tackled, so that the users appreciate why and when each step is taken.

There should be a formal procedure for identifying and investigating projects before any detailed design is attempted. Many jobs have gone astray because the bounds of a project have not been clearly defined at the outset. In some instances it may be easy to define the problem that requires tackling, but almost every system is potentially a subsystem of another, or itself has many subsystems and associated procedures. It is easy for a systems analyst to get pulled in a certain direction by the forceful personality of a line manager anxious to improve his department's efficiency. It is also necessary to monitor the work of individual systems analysts to ensure that the resources being expended on a particular project are commensurate with the benefits to be attained.

In the following text it is suggested that three distinct steps are observed in identifying and designing a system, and at each step the work is reviewed to check that it is proceeding in the way that management requires. These steps are:

(a) Initial identification of a problem and its related functions;
(b) Feasibility study
(c) Detailed specifications of the new system.

Fig. 9.1 illustrates the relationship of these functions in a live organization.

PROJECT SELECTION AND INITIAL INVESTIGATION

The object of the initial investigation is to examine an organization, or some unit therein, to see whether certain goals required by the management can be attained. Here it is assumed that a problem has been generally identified and a

CONTROL OF SYSTEMS DEVELOPMENT

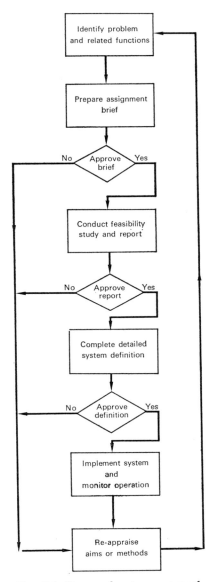

Fig. 9.1. Stages of system approval.

proposal has been made that the systems department should put forward a solution. At this stage it is not established whether data processing equipment will be necessary.

The first step is to see what are the specific goals required, and these should be expressed in a brief statement prepared by staff of the relevant operating department and the systems department. This statement should be agreed at the highest possible level, and must then be announced so that all departments concerned are encouraged to support the systems staff.

Quite often, in practice, a request for systems effort is received from an executive acting unilaterally to improve some aspect of his department's operation. He may be very precise in defining his requirements, but, on the other hand, may simply make some all embracing request – for example, 'to improve the service level provided by the workshops'.

If a study is started on such a vague request, the project will probably run into trouble at an early stage. Not the least of the difficulties will be concerned with knowing when to stop the initial investigation. However, it is not unusual for systems analysts to start their work with a brief of this sort; indeed, they often come up with sound assessments of the situation, which when presented to the operating management, provide justification for improvements in specific procedures. However, in creating such an assessment the analyst is really helping the management to set objectives.

If the systems department does accept a brief in this manner, it has to ensure that a balanced picture is presented to the management concerned. It is generally necessary to avoid accepting biased assignments from executives who are acting alone in some area where others in the organization should be consulted.

A Steering Committee

This condition can be avoided if the whole data processing function is planned from the outset as defined in Chapter 2; furthermore, the plan should be guided by a committee of senior executives representing the various divisions of the organization concerned. These people are not data processing specialists but are responsible for the various operating departments; and, in this specific role, are used to guide the data processing activity to obtain the overall goals of the company. Their task, in conjunction with the line management and data processing management, is to identify those areas of the organization where major benefits are to be obtained. They can call for specific studies to be made by systems staff and will evaluate the reports thus produced.

If it is known that a major system has not kept up with changes in the business of an organization, then it may be obviously a candidate for a detailed study. Sometimes the objective of the study is to simply discover why the current system is failing to produce the results required by manage-

CONTROL OF SYSTEMS DEVELOPMENT

ment or it may be that line managers have requested the study because they are not fully aware of how efficiently the existing system is operating. In each situation the analyst must be given performance criteria, and his job should be to see whether the system meets the requirements expected by management and, if not, to suggest possible ways of improving the system.

Projects authorized by the steering committee in this way are more readily given concerted support by operating managers as the investigation gets under way.

Developing the Assignment Brief

The objective of the initial study is to produce an agreed assignment brief which sets the general direction of the subsequent systems work; the aim here is to understand the current role and development of the departments concerned and to match their performance against company objectives in nominated areas.

The initial investigation sometimes begins with an analysis of the procedures already in use within the problem area. This first step can often get out of hand; the analyst may go on and on collecting more information and continue to document facts that he discovers about the organization and its people. There is certainly a knack in knowing what is important and what is not, but usually this activity can be kept in perspective by setting up project teams involving the relevant operating managers and members of their staff in aspects of the study. The object of a project team is to combine the experience of the operating departments and the systems people in identifying more closely those detailed procedures that need to be improved.

The initial study should attempt to answer the following questions:

(1) Is the current system producing useful results and is it related to the functional requirements of the organization?
(2) Have there been changes in the organization, its environment, or its policy, which necessitate system changes?
(3) Do management require more information from the system?
(4) Where the current system is inefficient, can this be expressed in meaningful terms, e.g. its cost per unit of work?
(5) Can the good characteristics of the current system be similarly expressed, e.g. its ability to meet sudden critical demands?
(6) Are there any technological factors that imply the need for change (e.g. improved methods for production of a commodity)?
(7) Can the objectives sought be obtained by minor or major changes to the new system?
(8) Are there alternatives to the current system and what are they?

The initial study will by now have indicated the aims of the planned system and may have indicated whether or not a computer or other data processing machinery is required. It may suggest methods by which the original aims can be achieved, and may specifically refer to certain desirable benefits to be obtained, e.g. quicker and more reliable information for operating departments, improved means of communication, better utilization of manpower or capital equipment, etc.

The assignment brief produced from the initial study is then presented to the steering committee who must review it to see that it covers the bounds intended and to take note of any further benefits or potential benefits afforded by the proposed system. The steering committee will now have an opportunity to consider whether their original request for the study remains pertinent. This stage is, therefore, the first opportunity to review the project, and if work is to continue the systems department should go ahead with a detailed feasibility report.

It is worth stressing once more the difference between the assignment brief and the feasibility report. In some cases the very nature of the problem may imply that the initial study will indicate the feasibility of achieving certain aims by using computer media. This is particularly true where the organization concerned has a tradition of data processing and has, say, already established a number of files which can be used to satisfy needs for information. The nature of the problem and the level of maturation of present file structures in the organization will therefore affect the nature of the initial study. The distinction should, however, be drawn as follows where any doubts are entertained:

(a) The initial study seeks to understand the problem and define its bounds; it may suggest the general way in which data processing resources can be directed to solve the problem, but essentially enables all potential users to specify their functional involvement;
(b) The feasibility study indicates whether a particular solution to a problem is viable, and outlines the nature of the proposed system, resources required when operational, its likely costs, and performance.

FEASIBILITY STUDIES

The feasibility study should be undertaken when the steering committee has authorized the assignment brief stemming from the initial study. The objective now is to see whether the initial proposals for a new system are valid, and to assess the costs of running the new system, the likely hardware and software resources needed, and the supporting facilities required by operating divisions of the company. This study should result in a more detailed report for presentation to the steering group or authorizing body. It should state the estimated time required to perform the detailed design of the various procedures and file structures and identify particular problems that may have to be overcome in this work or during implementation.

Within the report, the file structures will be defined in outline, and particular regard will be given to assessing the size of files and the volume of transactions to be handled in a given period.

The type of storage media necessary and the type of peripheral devices required must be given. Particular attention should be paid to items of equipment or associated facilities that are not currently available to the organization, especially those items which can be obtained with long lead times only. Data communications facilities, for example, often need to be planned many months in advance.

Other devices requiring careful consideration are those necessitating new technology or needing extensive software support. Data collection devices may need to be provided at many locations within the organization, and careful consideration of their operating environment may be implied, in which case some pilot scheme may be recommended by the analyst. The feasibility report enables the data processing manager to develop an equipment plan for his organization to meet the requirements of the many systems developing under his control. Not all feasibility reports will necessitate any major revision of the equipment plan, but all will have some effect on the production schedules of the computer operations branch.

In general, an equipment plan will be revised at certain key points in the evolution of the organization. Obviously, when a computer installation is first planned a great deal of forward thinking is undertaken, and the steering group and the data processing staff will be heavily engaged in assessing the feasibility of projects and developing a policy that will satisfy all major needs for perhaps three or four years ahead. Then the systems staff concentrate on the detailed design and implementation of the initial projects, and subsequent jobs tend to be considered in the light of hardware and software resources available at the time of their initial study. This tendency can inhibit forward planning for lengthy periods, and the data processing department should be careful to recognize the symptoms as they arise.

The operation of the steering committee and the sincere application of the principles laid down for conducting the initial investigation will help to ensure that all problems are considered strictly on the ability to meet the problems of line management.

Project Teams

Again it should be stressed that the feasibility study should be conducted with the involvement of the line management concerned. Although the feasibility report is in many respects a specialized document, reporting upon many echnical considerations, it should be developed by the systems analyst through project teams which enable the problems of line departments to be matched against the facilities needed to solve them. As the systems outline is developed, a series of presentations should be planned to get across the major principles of the design and to encourage users to visualize their problems in context of the proposed system. These presentations are used to introduce the principles of the system and pave the way for the final feasibility report.

Reviewing the Feasibility Report

The feasibility report has to be reviewed by the steering group; this provides the second major opportunity for management to assess the proposed system. The report itself may be against continuing with the outline given in the initial study, e.g. because the storage requirements of the system cannot be adequately met with the existing media. The report may suggest an approach whereby some of the original problems are tackled, but other problems in the assignment are left to a later stage when other software or hardware facilities are available.

It is for the steering committee to finally decide whether the project is to be continued or not. They are more able to do this if the systems analyst has produced a good document, setting down clearly the benefits that are obtainable by the implementation of the project, and suggesting a program for the later development of the project to meet other desirable aims. The report must deal fully and sincerely with all the problems raised by the initial assignment.

The steering committee may then agree to go ahead with the detailed specification of the system or may place the project in a queue with other proposals to await the authorization to go ahead at some future date. This situation is not as bad as it sounds; indeed, a steering committee cannot be expected to make intelligent use of the data processing resources until it is armed with a number of such proposals forming a plan for maybe four or five years ahead.

To the systems manager this approach enables him to plan his operation to meet a number of more or less known commitments. Not only does this enable him to plan his manpower and training requirements, but permits the systems analysts to design file structures and procedures with foresight of the problems to be encountered in the future. Perhaps more importantly, the data processing department are able to assess future hardware and software needs.

The depth of the feasibility study will obviously differ according to the nature of the applications being undertaken. A payroll system might be tackled as a problem in isolation, with fairly definite understanding of the nature and volumes of data to be processed, whereas the study of a number of applications in order to determine the general structure and nature of files forming a data base for a management information system would entail a far more detailed analysis at the beginning.

When tackling a complex proposal on the lines of the data base concept, the systems department is often faced with a difficult problem in communicating with the operating management and maybe with the steering committee. Specific applications treated in isolation can always be completed more swiftly and cheaply than an integrated network of systems. The advantages of the data base approach have therefore to be stressed at the inception of the data processing enterprise to all levels of management, and each subsequent proposal arising as the outcome of specific feasibility studies must be shown in context to the overall objectives.

If the systems department is to achieve any success at all, it must have the confidence of the operating management, and the adoption of a rigorous attitude in the selection of projects and in conducting feasibility studies is of the utmost importance. The development of the data processing applications can only be properly conducted with full participation of management in the organization.

DESIGNING THE SYSTEM IN DETAIL

When a feasibility report has been approved by the steering committee, the systems analyst has to design the system in depth and commit all the details to paper. This document is frequently called the *system definition*; not only must it define the operation of the new system but must specify, in detail, how the implementation is to be conducted and present a step-by-step plan for the creation of all files and the validation of the data used to set up the files initially. It serves as a reference document for all people associated with the development and future running of the project, including management and staff in the operating departments of the organization, the programming branch, and the computer operations branch of the data processing department.

To a large extent, many details of the system will have been set down in the original feasibility study, or incorporated into the working files of the systems department at that stage. The principles of the system will have been explained to the user departments during the feasibility study, and, subsequently, the analyst must develop the detailed clerical routines with the departments concerned so that a full understanding of their commitments to the system is achieved. It is usually preferable to establish more project teams during the phase in which the detail of the system is designed; these teams can be used to conduct the specification of certain key aspects of the system – one team for each activity.

For example, the first project team should reaffirm the type, format, and volume of all output required from the new project. The operating managers can then be brought straight back in to the heart of the problem by confirming what, how, and when they need information. Detailed print formats should be drawn up and formally approved, and a clear understanding has to be established that from this point on the formats are frozen. Changes after this point incur high development costs.

If the feasibility study has been held up pending an appropriate time to develop the system, the systems department must ensure that there is still common agreement about the performance and costs expected in the routine operation of the system. If any changes in these factors become apparent during the design phase, a report must be made back to the steering committee via the data processing manager.

MANAGEMENT CONTROL SYSTEMS DEPT.
INVENTORY CONTROL SYSTEMS (P/F 91)
Project Team 47

Objectives: (1) To ascertain the data preparation facilities needed to handle data for the daily order processing system.
(2) To create a schedule for the machine room embracing all functions concerned with this task, and to produce a time-table indicating when such facilities must be introduced.

Membership: H. V. Smith, Order Processing Dept. (Chairman).
I. J. Wilkinson, Systems Analyst (Secretary).
H. L. Wordsworth, Data Prep. Supervisor.
N. V. Rimmer, Stores Control Supervisor.

Reporting: A final report should be made to the Inventory Control Steering Party by 22nd May 1969.

MCS 23/3/69

FIG. 9.2. Terms of reference for a project team.

Project teams should not be allowed to continue for weeks at a time: each team should be given some definite objective and should be conducted to achieve a result by a nominated target date; see sample terms of reference in Fig. 9.2.

The membership of project teams must be decided according to the nature of the problem being investigated, each team being set up for a particular problem. There may need to be perhaps ten or twenty individual teams dealing with different aspects of the same systems project, and continuity can be provided by having certain members of the line departments concerned and members of the systems department on all teams. Final reports of project teams will be circulated to all managers of relevant line departments, the

CONTROL OF SYSTEMS DEVELOPMENT

management of the systems branch, and, where appropriate, to the steering committee.

The detailed design of the system must ultimately be the responsibility of the systems department, and a senior analyst should be given the final responsibility for producing a system. He must be able to plan the various activities, guide the work of the project teams, and direct the work of the systems and programming sections. He must be an excellent communicator, and be able to deal tactfully with all members of the organization. The systems manager and the managers of relevant line departments may have responsibility for the project, but are there mainly to assist the systems analyst, to ensure that he has access to information, to confirm that the departments are willing and able to provide necessary support, and to read reports and systems specifications for approval of the final application. During implementation, the systems analyst should continue to operate within the project team structure to ensure that resources are available for the successful conclusion of each activity.

In developing the final systems definition, all file and record formats, input formats, source documents, and codes will be specified and presented under the general headings given below:

(a) The aims of the system – with cross-references to the appropriate assignment brief or feasibility report, with benefits and relevant costs included;
(b) Description of the system – a brief description of the files maintained and reports obtained, with computer run-flowcharts accompanied by narrative;
(c) Source document specifications and outline clerical procedures;
(d) Implementation plan – program for file creation and validation of data;
(e) Equipment utilization – equipment required in routine operation of the system, and during file conversion and implementation;
(f) Detailed file specifications;
(g) Lists of codes;
(h) Output specifications;
(i) Program specifications.

A more detailed description of the content of this document is given later in Chapter 10 which deals with documentation in the systems department. Copies of the systems definition must be distributed to all managers of departments involved in the project and to the systems department manager, computer operations manager, data processing manager, and members of the steering group. Other authorities may be identified within the organization, e.g. the internal audit department or chief accounting authority. Formal acceptance of the system is then signified from those concerned by the signing of an acceptance sheet.

At this stage the steering group has the third opportunity to review the project to see whether it still contributes efficiently to the goals of the organization. Care should be exercised to see that this operation is not simply a question of rubber stamping the document. Obviously one cannot expect that senior executives on the steering group will find time to read all details of the specification, but an experienced steering group will soon find time to question their operating managers and the data processing management to establish whether the procedures are capable of meeting the goals required within the economic limits laid down originally. In the final analysis it is the responsibility of the data processing manager or the chief systems analyst to ensure that the systems definition is complete. When the systems definition is approved, the work of programming and implementing the system can begin.

PLANNING FOR SYSTEMS DEVELOPMENT

Projects Require Long Development Times

We have discussed a method for controlling the selection and development of systems projects and have indicated that the function of the steering committee is to authorize and monitor systems activity to ensure that it conforms with the overall aims of the company concerned. Many large systems take two or three years to develop and install – sometimes longer development times are experienced when many different departmental loyalties are encountered. Implementation costs and normal running costs are often high, and a system may need to run for, say, two years before the initial investment is repaid. In this atmosphere it is essential to work to long-term plans; any system installed should be expected to run for perhaps five or six years before major changes are necessary.

The establishment of an overall plan is particularly important when designing management information systems. In this type of application, research has to be conducted to ascertain the functions and needs of the organization and its individual units – both as they are today and as they are intended to be in five years' time. This implies that the systems must be developed in close cooperation with the corporate planning group within the organization as well as with senior executives in operating divisions.

Systems planning involves thinking about methods to assist other departments in attaining their particular objectives. This is often hampered by the fact that other departments do not themselves develop their own targets for more than perhaps a year or so ahead.

Breaking out of the Groove

The biggest stumbling-block to planning is that many managers of line departments consider planning a luxury that cannot be afforded. They are often too busy meeting day-to-day crises, and cannot condition themselves to think very far ahead. This attitude can create a vicious circle – activities get more frenetic as time passes and problems get worse.

Many departmental heads promise themselves that planning will start 'tomorrow' or 'when the current panic is over', but, of course, the happy day rarely arises. More problems occur and more short-term efforts are made to beat them off. Data processing departments themselves are just as prone to this type of behaviour and, where this is so, they lose a great opportunity to set an example to their users. If such attitudes are not controlled, the systems department may find itself aiming too low in its objectives.

Here we can see just how important the systems steering group can be – not only do they control the selection of projects as previously suggested, but must perforce make sure that a company plan exists and that its contents are thoroughly communicated down the management line. The steering committee thus becomes a jumping-off point for major policies, and is an important instrument in controlling the destiny of the organization. Once a viable plan is developed, the members of the steering group can see that their own staff are brought into the plan and make certain that they are encouraged to devote time to taking an active part in the design sessions that affect their own operations and development.

Major System Planning

Many of the applications developed for business organizations are of the management information type. At the beginning of such a development it is necessary to spend many months of research to understand the existing structure of the organization and to plot the flow of basic streams of data. The major events that give rise to information have to be visualized against this background of the individual systems and subsystems that constitute the total company system. With such an approach to systems planning, it is possible to visualize the file structures and systems necessary to meet the future shape of the organization.

Divisional representatives can soon see that it is necessary to be frank about their current problems and future development programs; how else can they expect to receive the data processing resources that they will require? The systems department must in return respond by undertaking a thorough feasibility study in each area requested by the steering group, and should be careful to present all relevant facts and details, impartially, in feasibility reports.

Feasibility studies will show the benefits and resource requirements of each project, and separate project studies may be necessary to consider the interface areas between major systems. From these studies the systems department will prepare an equipment plan and a software plan.

SYSTEMS ANALYSIS IN BUSINESS

Reviewing the Plan

Most organizations start with such a plan when initially setting out to install their first computer, but often these plans are not reviewed frequently enough. In consequence, the organization becomes slow to react in changing circumstances; sometimes, as a result, the company is critically short of data processing power when, say, their marketing activities require expansion or some technological breakthrough affects production systems in their particular industry.

The plan must be kept under review both at the steering level and at the working project level. Each project should be planned to have effect in reaching the total objective, and any departures from the original assignment brief must be considered by the steering committee. New business situations are also to be examined to see how they will affect the current plan – the various divisional representatives should bring to the steering group any problems that arise in this way. For example, the company may launch a completely new product or service, it may take over another company or decide to situate some of its production resources in a distant location. Most events of this nature can be anticipated many months in advance, and every effort should be made to make plans to meet the situation; even a year may not be long enough if new equipment having a long lead time is required.

Symptoms of Bad Planning

Where only lip service is paid to long-term planning, it may be apparent that the systems department has a tendency to attempt to solve all problems in terms of the equipment and software currently available, with the consequence that new techniques are not brought into the organization. For example, improved methods of data collection and transmission may be passed over, or problems requiring a fast response may be tackled inappropriately in batch processing mode. By following such a path, there is usually a failure to make timely equipment decisions, and critical shortages of computing power and storage facilities may be evident. The training and development of computer staff is sometimes stunted thereby, and when the decision is finally taken to break out it is probable that the staff will be inadequately prepared.

Planning ahead for several years will usually make more apparent any weaknesses in the current organization, both in line departments and in the data processing unit. Key executives are able to discuss these problems and to see them in the light of the total company plan. The development of this overall system plan, as in all systems work, has to be done in participation with management, thus promoting good internal communication and serving to identify the overall goals of the company. In this way the organization is able to adopt a posture that enables it to condition itself for coming events.

CONTROL OF SYSTEMS DEVELOPMENT

In the absence of a long-term strategy, the systems department will be beset by many changes of direction that will send costs soaring and perhaps lower morale. The various users will be unable to identify themselves with an overall strategy and, in consequence, will continue to pull in different directions. When a systems department finds that it is being continually committed to produce systems for short-term problems, then clearly there is not a good company plan in existence, or it is not being effectively promoted and managed.

Need for Clear Authority

There is often a general diffusion of responsibility where procedures cross departmental or divisional boundaries; and, where there is no defined authority for providing general management information within a company, one often finds that ambiguous and irreconcilable facts are put forward by different parts of the organization. The sales department and the production department may each produce different figures as representing the company order book, whereas the commercial manager may show yet another set. This situation will usually arise because there has been no consistent plan, and each department has developed its own records and procedures to provide data for running its own operation. Perhaps each may have a different definition of an order.

Where there is a broad-based plan for systems development, with all parties contributing, there is the ability to establish future standards of operation throughout the organization, and policy differences can be identified and settled in laying down the procedures for particular systems. Only a sincere commitment to this attitude will ensure that the systems development is conducted to solve the daunting communication problems that exist in large organizations.

CONTROL OF PROGRAMMING

Importance of the Systems Definition

To some extent the programming function is one in which it is possible to make fairly accurate estimates of the development costs and the overall time needed to complete a project. If an accurate and detailed systems definition has been produced in the first place, the programming manager will be able to schedule and control his staff in an efficient manner. The first job in tackling a project will be to make a study of the systems definition to ensure that the work has been adequately specified.

Work entering the programming department can be considered as falling into two main categories:

(a) Programs that are to be used routinely as part of a system and which have been specified as part of a systems definition;

(b) *Ad hoc* programs to do one-off jobs, e.g. editing runs in file creation.

The work in the first category can be anticipated and budgeted for some way ahead, but work in the second category often arises at short notice, and the programming department may have difficulty in providing the necessary manpower. Some organizations approach this problem by keeping programmers available to tackle this *ad hoc* work separately within the programming department. Other organizations attach programmers to systems development teams, or expect the systems analysts to carry out this function.

Wherever *ad hoc* tasks are tackled in the programming department, there will be a need to specify such programs in detail as part of the systems definition or as a supplement to it. In practice, one finds that problems arise in system implementation that require a fairly speedy response, and the need to produce detailed documentation and schedule the programming as part of the overall programming load becomes impracticable.

On the whole, it is better to form compact teams in which systems and programming skills are combined. In some organizations this principle is followed to the extent that all programs, *ad hoc* or otherwise, for implementing and running a particular system are developed as part of such a team. This has the advantage that communication between the analyst and the programmer is simplified, but this should not mean that systems and programming documentation are avoided completely. The advantage of this policy is most apparent where programming work entails mainly suites of batch processing programs for the functions of input transcription, editing, file updating, and reporting, particularly where generalized standard software has been developed to meet these basic functions as described below.

It has been said that systems should undergo a continual appraisal and modification if they are to be kept in line with the aspirations of management. Any programmer who has sat down with a neat and tidy specification will tell you that, in fact, changes are forthcoming from the moment that the analyst hands over the document. Changes to file layouts or program procedures can play havoc with work schedules in the programming department, and it is advisable for the programming manager to curb this tendency by applying a freeze on the specification when it is handed over, requiring all amendments to be raised formally by the analyst in writing. In this way unimportant changes are avoided, and amendments can be considered against the relevant schedules.

Standardized Routines

Most computer manufacturers supply utility software and standard subroutines with their equipment to perform certain basic operations, e.g. to

CONTROL OF SYSTEMS DEVELOPMENT

perform conversion of a character field to an equivalent binary number. The subroutines can thus be employed by programmers in their programs, and are parameterized to perform particular tasks required. It is an advantage to develop more routines of this nature for other operations that arise often in the work of the department, so that the same work is not continually duplicated. Where this is done, most jobs can be more or less visualized as consisting of a number of proven subroutines.

Consider for the moment any printing program that prints data from a magnetic tape file; it can be broken down into a number of operations that recur in all programs of this type:

(1) Opening files and setting switches to predetermine initial conditions;
(2) Printing control characters to align preprinted stationery;
(3) Setting up data for heading lines;
(4) Transferring heading lines to output buffer where necessary;
(5) Reading input records;
(6) Editing records to output buffer;
(7) Printing detail lines;
(8) Controlling page overflow (perhaps with sub-totalling);
(9) Checking keys of records for control breaks;
(10) Printing totals at control breaks.

These features can be written generally into one basic program consisting of a number of subroutines that are parameterized by the user to do particular functions, but which are, in most respects, capable of operating on any data formats. The principles of this approach can be applied to many other basic processing operations including:

Input transcription programs;
Validation programs;
Editing or indexing routines;
File updating programs;
Reporting programs.

Software of this type is known as data management software. Its purpose is to provide systems analysts and programmers with a language for getting programs written and tested quickly and efficiently, each program being made up of standard subroutines combined together and parameterized to do specific tasks.

Estimating and Scheduling Programming Activity

The systems development function is a service to the organization as a whole, and many operational developments are dependent upon the efficiency with which the systems and programming staff are able to perform their function.

163

SYSTEMS ANALYSIS IN BUSINESS

The scheduling and implementation of many costly systems projects is conditioned by the capability of the programming department. It is, therefore, necessary to estimate work loads with accuracy and to plan jobs in such a manner that realistic budgets and priorities can be established.

To estimate and schedule programming activity, standards have to be developed for programmers of a given grade performing jobs of a particular nature. The standards can be developed by measuring activities over a length of time to obtain norms for conducting elements of each job. The standards have to be understood and accepted as realistic by the analyst/programmers, and there should be a suitable procedure for reviewing and refining them.

Programming is generally considered to be a creative activity in which a great deal of original thinking is necessary. This leads to the conclusion that job estimating and scheduling is difficult and that the work content of any particular job will be unpredictable until the job has been planned in some detail. It is true that some programs are of a very specialized nature, but in many commercial organizations most programs are made up of routines which are similar to routines written and tested previously; programs can thus be classified into categories and data collected over a period of time to record performance for each category.

Programmers themselves vary in their capability, e.g. some are more suited to scientific applications than others, or have been trained in particular techniques. Their experience and training is relevant to their ability to develop programs of a particular nature. Realistic estimates of the time and resources need for a particular task can only be made where adequate training and grading schemes are provided.

The programming function can be considered to start upon the completion and acceptance of the systems definition by the leader of the programming team. Changes to this definition should be discouraged after acceptance. A study of the systems definition has to be made before formal acceptance in order to evaluate whether the specifications are complete; test data and expected results should be provided by the analyst at this time.

The successful completion of a programming task occurs when it has been fully tested to provide the desired results from the test data and when documentation of the programs has been finished to agreed standards.

On accepting a systems definition, the leader of the programming group has to break the whole task down to a number of integral activities which can be estimated separately. Each activity is assessed individually, and the overall time and resources needed are derived by aggregating the individual elements.

The nature of these activities depends to some extent upon the programming language used and the organization of the programming function. Assuming that all programming is conducted using a basic assembly language, the programming load can include the following activities for each program.

CONTROL OF SYSTEMS DEVELOPMENT

These activities may be carried out by one programmer or by a number of people of different grades:

(a) Analyse the systems definition to see if it is complete and to identify individual programs;
(b) Evaluate individual program specifications and prepare outline flowcharts breaking each program into segments where necessary;
(c) Check outline flowcharts;
(d) Prepare detailed flowchart of each segment and specify also interface between each segment;
(e) Check all detailed flowcharts by dry running through the logic on paper with test data;
(f) Code and check segments of programs;
(g) Prepare test data for checking programming logic;
(h) Punch and verify programs;
(i) Compile and test programs using programmers test data;
(j) Analyse test results, identify and debug errors, re-compiling as necessary to obtain correct programs;
(k) Test programs as a complete suite using analyst's test data for the system;
(l) Complete documentation and prepare operating instructions.

These activities are more or less independent, but the work of documentation will be accomplished as the job is progressed, and estimates for each activity should, therefore, include time needed for completing and filing relevant documents to agreed standards.

The availability of computer time is one function that can effect program development schedules, and it is necessary, therefore, to evaluate the time needed at each stage so that the computer operations manager can arrange machine-room schedules to encompass these requirements. For a given type of job and grade of programmer, an allowance of time for each of the following functions should be made:

(a) Compilation and listing;
(b) Preparation of test files;
(c) Program and segment testing;
(d) System testing.

Trainee programmers have to be carefully supervised to see that they are within agreed standards for these operations. If a programmer consistently falls outside the developed standards, then he is probably in need of guidance with some aspect of his work.

Programming Documentation File

The next chapter deals in some detail with systems documentation and in so doing dealt only lightly with the needs of the programming function. In the following text, this aspect is developed in greater depth, and relates particularly to the documentation standards necessary when using a basic assembly language. These standards could be adopted in writing both commercial programs or substantial items of software. The underlying assumption in the following text is that the programmer has received a program specification from the analyst as part of a systems definition, and the documentation file described below will be created in developing the detailed procedures from the analyst's specification.

1. *Title page and description.* It is pointless to give a lengthy description of the program if it is already covered in the systems documentation. Nevertheless, the program file will be used at some time in the future, when it may have to stand in its own right. Therefore certain information should be provided at the beginning of the file to identify the program. This description will seldom need to exceed one or two pages, but should state the program name, number, say what the program does, in what system it is used, and its basic inputs and outputs should be identified. It is also essential to indicate how frequently the program will be used daily, weekly, monthly, etc. The name of the programmer responsible for the development of the program and its associated documentation should be given.

2. *Program specification.* This section can simply be the original specification provided by the analyst and accepted by the programming section. It is filed for convenience with the program documentation file so that cross-reference can be made easily to file layouts, output charts, etc. It should include the program description, print-out specifications, and relevant file specifications from the systems definition.

3. *Outline flowchart.* The outline flowchart (see previous sample in Fig. 1.2) is used to indicate overall structure of the program; it should be accompanied by a narrative which amplifies details shown on the chart. The two combined must identify and describe the major segments of each program and the routines within them, show input and output operations, specify links between segments, and all entry points and halts. A separate list of data elements with details of format and size should be given where such elements are used by more than one segment. A separate record (Fig. 9.3) should also be made giving details of switches used to control branching between segments or routines drawn on the outline flowchart; similar records should be provided for counters specified as part of the control logic at this stage.

4. *Detailed flowcharts.* Detailed flowcharts are needed to break down the major routines produced at the outline stage to the detailed logical steps

CONTROL OF SYSTEMS DEVELOPMENT

necessary to each major function. Detail charts with narrative are often developed independently for each segment and in sufficient detail to permit coding to be completed direct from them. Records of switches and counts

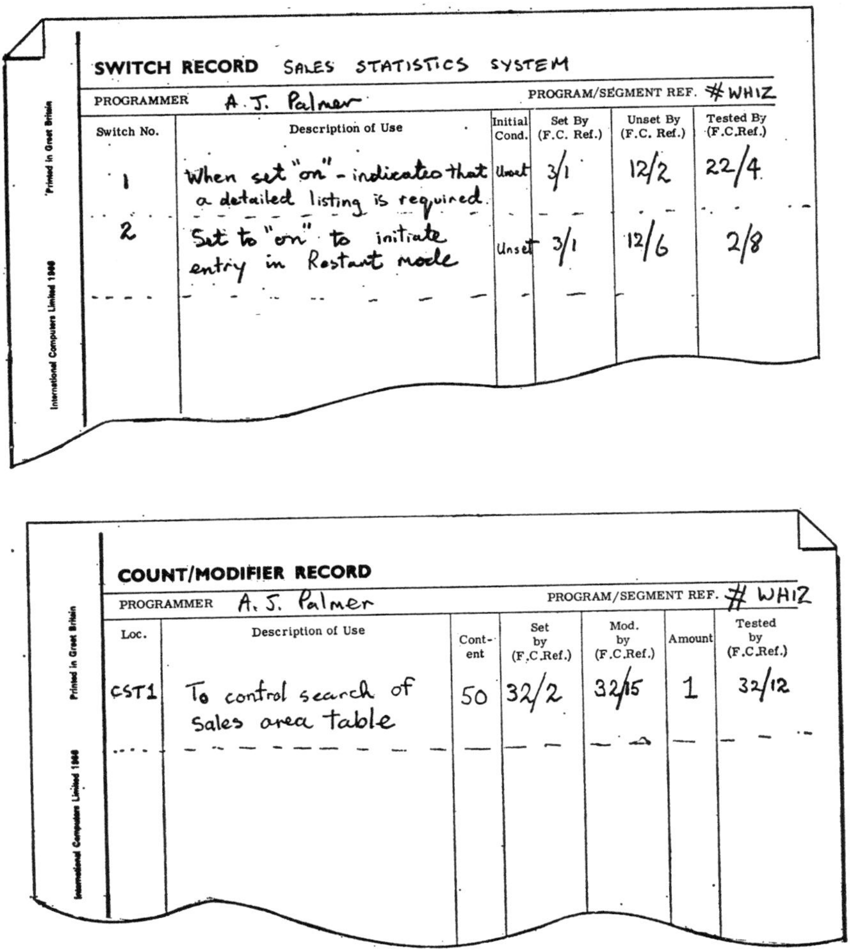

SWITCH RECORD	SALES STATISTICS SYSTEM					
PROGRAMMER	A. J. Palmer			PROGRAM/SEGMENT REF. # WHIZ		
Switch No.	Description of Use		Initial Cond.	Set By (F.C. Ref.)	Unset By (F.C. Ref.)	Tested By (F.C.Ref.)
1	When set "on" - indicates that a detailed listing is required.		Unset	3/1	12/2	22/4
2	Set to "on" to initiate entry in Restart mode		Unset	3/1	12/6	2/8

COUNT/MODIFIER RECORD							
PROGRAMMER	A. J. Palmer			PROGRAM/SEGMENT REF. # WHIZ			
Loc.	Description of Use		Content	Set by (F.C.Ref.)	Mod. by (F.C.Ref.)	Amount	Tested by (F.C.Ref.)
CST1	To control search of Sales area table		50	32/2	32/15	1	32/12

FIG. 9.3. Examples of documents used to record details of programs.

used in the detailed logic design are completed as before, and any complex logic in the flowchart should be explained in the text.

5. *System test data.* The system test data will be provided by the analyst to test out the program when used in the context of the overall suite of runs. It

should include details of data to be generated as test files, listings of specimen input data used to update such files, and also must specify the results expected.

6. *Programmers test data* A programmer will often find it necessary to develop test data for checking out the detailed logic of certain program segments, particularly where he has used a programming technique calling for complex logic. The programmer should satisfy himself that all branches of the program are adequately tested, and he will usually have to develop his own test data to do this.

7. *Coding sheets.* Coding sheets should be filed after being punched and verified; these can be replaced by a compilation listing when the particular segment or program has been compiled and tested out.

8. *Test results.* Test results will include report samples, error listings, and file dumps produced using the test data.

9. *Operating instructions.* Operating details relating to all run-time conditions must be specified, either in detail or in note form, for the analyst to generate final operating documentation, as previously described.

Chapter 10

Documentation in the Systems Department

THE NEED TO COMMUNICATE

In the opening of Chapter 9 an approach to the problems of project identification and evaluation was given. This is one of the first tasks undertaken in getting a project off the ground, and from the outset we can see that there is a need for the analyst to have the ability to write effectively. The original *assignment brief* has to be concise, perhaps not more than one page, and yet describe the aims and bounds of a project in sufficient detail to meet the approval of senior management. The feasibility investigation which follows the approval of a major assignment generates the most important document of all; it has to develop and expand the nature of the problem, suggest likely ways of tackling the job with an evaluation of the costs and benefits in each case, and make recommendations and an outline program for developing and implementing the project. The *feasibility report* has the contradictory aim of getting across a fair amount of detail; yet it has to be capable of communicating effectively to senior executives who usually have little time to spare. The organization and presentation of the material has, therefore, to be of a high standard.

When a project is approved for the detailed design to begin, the major task is to develop the *systems definition*: this document can have many purposes and is best organized into a number of sections aimed at those concerned, e.g.:

  Management;
  Programmers;
  Computer operations staff;
  Auditors;
  Clerical supervisors.

This list can get too long, and it is a good premise to try to limit the aims of each document, and we shall have more to say on this score later. However, it can be stated at this point that the purpose of the document is to describe the details of the programs, procedures, and documentation needed in running the system; in practice it is often necessary to limit the descriptions of clerical procedures to outlines that are later developed in detail elsewhere.

Systems projects usually embrace a number of functions within an organization, and there are many people involved in the live running of a system. Each person must know what is required of them and when. It is the systems analyst who has the final responsibility for seeing that documentation is produced to specify these requirements; and he will have to prepare a careful plan for the education for all those connected with live running of the procedures. For this purpose he must be able to write clear and concise procedure manuals – or supervise the preparation of such documents.

The *operating instructions* for the computer procedures have also to be documented along with procedures for preparing input media and for dismantling the job and distributing the output.

Whereas it is universally accepted that documentation is necessary, it is important to see that the work is not allowed to dominate the main tasks. Documentation can represent a heavy load, and, if possible, it should be shared by members of the systems team. Where documentation is inadequately prepared and maintained, it is usually because the systems staff are not supported with reasonable and effective standards. In some companies the systems department may have a supporting section for developing procedure manuals, particularly those relating to operations outside the immediate computer procedures; in other establishments the clerical supervisors may be encouraged to develop manuals for their own staff based on outlines in the systems definition.

DOCUMENTATION STANDARDS

The specification and documentation of the computer procedures and data files is most important, and should not be jeopardized by an undue emphasis on other matters in the documentation standards of the department. Analysts and programmers should be provided with a standards manual developed for use in their particular environment. Such a manual should identify the essential elements of each document and provide a common format to each item of the overall documentation without being unduly restrictive.

Some organizations choose to have pages of their documentation prepared on preprinted forms or charts; indeed, computer manufacturers sometimes recommend this and provide samples for their customers. The essential point is simply to ensure that specifications have a standard content and method of organization and a predetermined format for title pages and contents lists.

Documentation is needed to maintain continuity in the absence of the analyst; it safeguards the systems manager should the analyst leave the company concerned. Also good documentation permits members of the systems department to communicate effectively with one another and improve mobility within the department. Perhaps the most important standards are those relating to the organization of documents in working files; if this aspect is dealt with sensibly, projects will not be so heavily dependent upon individual analysts. In the next sections some recommendations are made for the establishment of standards for systems documentation. They are based on standards used in a department which employs analysts to carry out specific tasks; programming is done by the analysts using a high level language suitable for commercial and business applications. This language is to a large extent self-documenting and, therefore, very little program documentation is needed.

FEASIBILITY REPORTS

A feasibility report is produced to show potential methods of solving a particular business problem. It should be developed to define the nature of the

---

Part 1 – Problem Definition and Recommendations
    Section 1 – Introduction
    Section 2 – Problem definition
    Section 3 – Possible solutions
    Section 4 – System evaluation
    Section 5 – Summary and recommendations

Part 2 – Computer System Details
    Section 6 – Computer processing structure
    Section 7 – Nature of output
    Section 8 – Input data
    Section 9 – Computer files
    Section 10 – Programming estimates
    Section 11 – Implementation plan and schedule
    Section 12 – Special equipment

FIG. 10.1. Contents of feasibility report.

problem and, wherever possible, seek to demonstrate the extent of the problem, e.g. its effects upon profits or service to customers. The ultimate aim is to produce recommendations for dealing with the situation, and to support these recommendations with assessments of the benefits and costs for running and developing any new system that may be necessary.

These assessments must be completed accurately, and this implies that the outline of the new system must be developed in some depth in order to evaluate all the relevant factors properly. Suggestions for the content and structure of the feasibility report are given in Fig. 10.1.

The feasibility report can be considered in two parts; first, sections which are aimed generally at company management and the steering group; second, details needed by the data processing management to plan their own activities. This is reflected in the contents given in Fig. 10.1, but this structure should not be rigidly adhered to if the circumstances do not fit. For example, it may be necessary to include a special section where a user department has to provide some complex and/or expensive supporting service in operating the proposed system. Let us now consider in more detail these sections.

The introduction should be a brief statement of the problem and refer to an initial assignment brief or any previous report that preceded this particular study.

The section entitled Problem definition is important; it is here that the analyst has to indicate the nature and bounds of the problem. It may, perhaps, be simply expressed, but, on the other hand, may be a comprehensive matter affecting the complete operation of a company or one of its major operating divisions. Those who come after in the project will be grateful for care taken at this stage. Also, it is wise to get approval of the draft of this section with the relevant managers as early as possible in the development of the study.

Under the heading Possible solutions the analyst has to describe the outline of a system that will solve the problem. Depending upon the nature of the problem, it might be pertinent to render more than one solution. For example, some problems may be satisfied by a simple batch processing system yet have potentially greater benefits where facilities are provided giving users access to data on-line. Both solutions could be presented showing the advantages of running each system and a comparison of the resources and time scales needed for each implementation.

The system evaluation is the section where the analyst must demonstrate the economic justification of his proposals. He must identify the advantages that the new system will provide and the resources needed for implementing and running the system. The advantages of the proposals can usually be classified under some of the following headings:

(a) Improved efficiency for the user departments concerned;
(b) Better service to users or customers;
(c) Improved quality of information;
(d) Reduction in overhead expenditure;
(e) Reduction in stocks;

DOCUMENTATION IN THE SYSTEMS DEPARTMENT

(f) Improved profitability;
(g) Improved communication.

All the claimed advantages should be quantified and supported by figures to demonstrate how the benefits are obtained. If necessary, pilot schemes should be proposed to test out the major principles of the system.

Against the advantages, the costs of the system should be shown including the factors shown in Fig. 10.2.

---

1. *Development Cost*
    (a) Manpower
        Clerical support
        Programming
        System design
        Data preparation
    (b) Computer time
    (c) Other services
2. *Running Costs*
    (a) Manpower
        Data preparation
        Data control
        Program maintenance
        Clerical support
    (b) Equipment costs
        Hire charges
        Maintenance charges
    (c) Stationery
    (d) Computer input/output or storage media

FIG. 10.2. Costs of systems development.

The section headed Summary and Recommendations is intended to provide a concise summary of other sections of the feasibility report. It must provide a practical assessment of the proposals and be suitable for communicating with the managers who are concerned with policy in the problem area. It may cross-refer to any other sections, where necessary, but should aim at being a self-contained statement giving guidance to the authorizing body.

If the system is to be installed as a number of phases, a schedule of these must be given showing the lead time for each phase and the project as a whole.

Part 2 of the report must be developed in sufficient detail to identify the hardware and software resources necessary for running the project. The problem here is to judge accurately the extent to which the detailed design work is necessary. The basic objective is to be able to assess whether any new equipment or major items of software are needed or whether the programming load can be handled with current resources.

The Computer Processing Structure should contain a general run-flow-chart of the procedures needed to update and maintain files in the new system. This structure may be provisional but must, at least, indicate the number and type of programs needed, so that programming loads and computer running times can be estimated. A narrative underlining salient points should be given.

The Output provided by the system should be specified in terms of medium, format, and volume. These need not be designed in detail but should have been considered in sufficient depth to be satisfied of the efficacy of the medium employed and the eventual uses of the output. The nature of the response required by the users must be described and the data elements have to be identified. Samples of printer outputs or visual displays, etc., can be included; the most important thing, however, is to identify the functional requirements of users and to establish the feasibility of meeting their needs.

The Input Data section should identify all principal streams of transaction data that enter the system and the data elements contained in such transactions. The methods of data collection or preparation must be specified, and the volume of the transactions and their source should be described. Any special data collection equipment to be used (e.g. OCR readers) must be identified and, where necessary, a separate section should be included to show performance details of such equipment. Any particular problems in data gathering must be presented, and proposals for conducting pilot schemes outlined.

The section headed Computer Files is intended to indicate the effects of the new system upon existing file structures and give outline specifications of new files to be established. Where existing files have to be amended, effects upon these files in terms of record formats, record size, and overall file volumes should be stated. In addition, the effects of these changes upon other programs which use the files have to be considered. For all new files the data formats, record size, and file volumes have to be defined and the file structure can be developed here, along with the sample layouts if necessary. The main objective is to identify the file media and storage capacity necessary.

In the next section, Programming Estimates should be given to indicate the nature of the resources required in setting up and implementing the system. The nature of the programs must be specified to identify complex or unusual programs; the requirements can be considered under the following headings:

(a) New programs of a relatively simple nature or relating to previous experience;

(b) Existing programs that require amendment to interface with the new system;
(c) Major software items, or complex programs, not previously experienced.

Some attempt must be made to assess the number of man weeks' programming effort needed to get this work completed; items (a) and (b) above should be easy to estimate, but item (c) may require an outline specification to be produced, followed by an estimate emanating from a specialized software development group.

The Implementation Plan needs to identify the major events to be undertaken in getting the system operational assuming that the feasibility report is approved for development. This should include an outline schedule incorporating programming, activity, file creation, equipment procurement and proving, education plans, obtaining documentation and stationery for running the system, and an outline of the methods to be used for proving and installing the system.

The section on Special Equipment will be needed only if new hardware techniques are being introduced into the organization: e.g. OCR devices. The reliability of the equipment has to be evaluated, and performance and operational characteristics checked out thoroughly. This may entail visits to other users to determine their achievements with the equipment concerned, perhaps arranging for stringent tests to be carried out by the supplier under specified conditions. The readers of the feasibility report will want to see summarized results of all tests conducted, with a report indicating why the equipment has been proposed.

WORKING PAPERS

Much of the documentation that the analyst has to produce can be developed from working papers that contain details committed to file at various stages in the investigation and design phases. The working papers can be considered as two distinct parts, each subdivided into a number of sections:

(a) The investigation file – papers relating to the current system;
(b) The system file – papers relating to the new system being developed.

The extent to which these papers are developed depends upon the nature of the problem and perhaps upon the accuracy and availability of documentation already existing. Documentation can become an obsession, and a surfeit of paperwork has to be avoided. Nevertheless, if working documents are correctly organized, the problems of producing final documentation for other purposes are greatly simplified.

The Investigation File

Section 1   Identification of the problem area and of the various systems to be reviewed.
Section 2   Identification of the departments concerned, their functions, and terms of reference in relation to the problem. Organization charts and a brief directory of people concerned.
Section 3   Flowcharts of the systems concerned with supporting narrative when necessary. Reference to existing documentation – procedure manuals, etc.
Section 4   Copies and descriptions of all forms and documents used in the system.
Section 5   Specification of the uses and structure of any existing files, whether manual or mechanized.
Section 6   Notes developed from interviews with various members of the organization in identifying current problems and methods of operation.
Section 7   Suggestions for individual features to be incorporated in any new system; identifying the source of any such proposals.

FIG. 10.3. Check-list for the investigation file.

System Design File

Section 1   A copy of the original assignment and relevant supporting papers and correspondence.
Section 2   A schedule identifying significant stages of the project with target dates and actual completion dates. This section will be revised, as necessary, to reflect progress and objectives.
Section 3   Sample output formats showing output content, format, frequency response needed, and details of distribution.
Section 4   File layouts, data element specifications, and code lists.
Section 5   Specification of computer run structure including run-flowcharts and narrative of actions required in each run.
Section 6   System test data.
Section 7   Program development details including details of source programs, program lists, program testing, test results, parameter specifications, operating instructions.
Section 8   Separate files containing minutes of project teams set up in developing the assignment.

FIG. 10.4. Check-list for the system design file.

DOCUMENTATION IN THE SYSTEMS DEPARTMENT

The investigation file stems from the activities described in Chapter 2 and is intended as a record of the work done in investigating and analysing the existing system. A check-list giving suggested contents for this section of the working file is given in Fig. 10.3.

The system design file is opened when an assignment brief is requested, and thereafter it is developed to record the analyst's thoughts on various aspects of the task; gradually the file is refined as the different elements of the system are accepted. A check-list is provided in Fig. 10.4 showing an outline structure of this file. Section 7 is included here since it is assumed in the example that all program development is conducted by the systems team using a high level language that is self-documenting.

THE SYSTEMS DEFINITION

The systems definition is developed from the working files to reflect the final structure of the system. The main purpose of the document is to obtain the approval of those persons responsible for authorizing the system and to act as a vehicle for communicating some detailed aspects of the system (e.g. program specifications). Some people prefer this document to contain every last detail of the clerical and computer procedures forming the system as a whole, but this tends to make the document unreadable for most recipients, and also consumes much unnecessary time in writing and printing.

If the working files have been correctly developed and maintained, they will provide sufficient details within the systems department for amending or comprehending any details of the computer procedures at a later stage. However, where there is a separate programming department it will be necessary to resort to a full specification of all computer programs in the definition itself so that the programming authorization can be obtained.

Approval of the supporting procedures necessary within the line departments is usually obtained in design sessions conducted in project teams which will have recorded and accepted details of new documentation and related procedures. The systems definition therefore will summarize these details with outline descriptions supported by documentation samples.

All systems definitions should be developed around a basic structure, and certain pages should have a strictly standard presentation, e.g. title-page, amendment record sheet, contents list. A suggested structure for the systems definition is given in the following text.

Structure of a Systems Definition

Section 1. The *preliminary pages* including a title page, contents list, and
  amendment record: an outline implementation schedule should also be in
  this section.

Section 2. A *glossary* of all terms that are related to the particular business area and which may be considered to be unfamiliar to the reader.

Section 3. A section entitled *aims of the system* which identifies the departments and procedures tackled in the assignment, the main files to be developed with a summary of benefits and costs. This section should be brief, one or two pages only, borrowing from the original assignment brief and feasibility report and cross-referring to these documents if necessary.

Section 4. The *systems description* is based upon details in the feasibility report if one has been produced. It should include a run flowchart and accompanying narrative.

Section 5. Samples of all *source documents* should be given in this section with an outline description of their use. Reference should be made to separate procedure manuals where these have been produced.

Section 6. *File specifications* can be developed from working files. Under this heading we can include all punched-card or paper tape files used as input, and all intermediate or main files on magnetic tape or discs. Pictures of file layouts are easiest to comprehend, but these should be supported by lists specifying structures of individual data elements and codes.

Section 7. *Output specifications* should be next; they can be finalized from the working papers or feasibility report with details of content, format, frequency, and distribution details. All totals given at control breaks should be specified with maximum and minimum sizes of fields, and vertical and horizontal line spacing. Any special stationery required must be specified.

Section 8. *The implementation procedures* should be described to identify all the major events which must be completed to get the system running. Particularly, it should specify the procedures necessary to create files, laying emphasis on important matters, e.g. sources of initial data, validation, coding, and editing of data. Peak loads for data preparation, *ad hoc* programming, or computer time should be identified so that resources can be organized to implement significant stages.

Section 9. A section headed *equipment utilization* should be included to identify the equipment resources needed to run the system when it is operational and to set up files initially. Such resources may include data preparation equipment as well as computer hardware. These time estimates are fairly easy to make for daily, weekly and monthly routines, and can be done with reference to run-flowcharts incorporated in earlier sections.

Section 10. The *program specifications* are probably the most important, and this section must be completed in detail if it is to be used for instruction to programming staff. It must be updated even for relatively minor changes and will need to be maintained throughout the development and operational life of the system. All processing operations to take place in each program must be given including arithmetic and logical operations upon data, validity checks, error detection, error reporting, control totals, opening files, closing files, dump and restart routines, run time options, operator messages, etc.

OPERATING DOCUMENTATION

Operating documentation can be considered in two categories:
(a) Instructions required to prepare all inputs and distribute outputs – referred to below as an Operations Guide.
(b) Instructions required to operate the job in the computer room – referred to as Operating Instructions.

A job may be a series of runs which are an integral part of a system and are run together to produce final results or some intermediate output. For example, a system may have a daily run, a weekly run, a monthly run, and so on; each of which will have separate operating instructions.

Operations Guide

This document can be divided into six sections as described below.

Job identification and description. A title-page giving job name and number plus a brief description of the job including a run-flow chart, indicating what the job does and its relation to other jobs.

Input and output. Identifies all data sources and nominates the relevant data control authority, and describes when and how communications are conducted. Schedules for collecting and transporting data are given. Similarly, all outputs (e.g. intermediate or final results, error reports) have to be identified and methods and times of distribution must be given.

Specification of run-time options. In many cases there will be conditions that can be set at run-time to change operations during the run, e.g. switch settings, parameter changes. These can be described here to enable a run-time sheet to be prepared as required.

Control of files. This section will identify all files used by the job and include instructions for generation control and the general security of each file.

Quality control. Provides details of checks to be made after output is produced or at some intermediate control stage to verify that operations have been correctly performed.

Apart from the sections described above it may also be helpful to provide layouts of main files used in the run. This may assist in carrying out simple *ad hoc* work performed by operations staff in support of the system. For example, sorting a file into a special sequence prior to a particular reporting operation.

SYSTEMS ANALYSIS IN BUSINESS

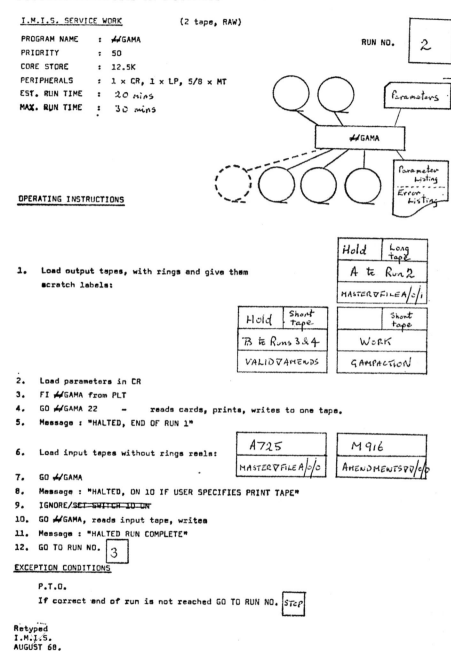

I.M.I.S. SERVICE WORK   (2 tape, RAW)

PROGRAM NAME   : ⊬GAMA
PRIORITY   : 50
CORE STORE   : 12.5K
PERIPHERALS   : 1 x CR, 1 x LP, 5/8 x MT
EST. RUN TIME   : 20 mins
MAX. RUN TIME   : 30 mins

RUN NO. 2

OPERATING INSTRUCTIONS

1. Load output tapes, with rings and give them scratch labels:

   Hold | Long Tape
   A to Run 2
   MASTER ∇ FILE A/c/1

   Hold | Short tape
   B to Runs 3 & 4
   VALID ∇ AMENDS

   Short tape
   WORK
   GAMPACTION

2. Load parameters in CR
3. FI ⊬GAMA from PLT
4. GO ⊬GAMA 22   —   reads cards, prints, writes to one tape.
5. Message : "HALTED, END OF RUN 1"

6. Load input tapes without rings reels:

   A725
   MASTER ∇ FILE A/c/o

   M916
   AMENDMENTS ∇∇/c/o

7. GO ⊬GAMA
8. Message : "HALTED, ON 10 IF USER SPECIFIES PRINT TAPE"
9. IGNORE/~~SET SWITCH 10 ON~~
10. GO ⊬GAMA, reads input tape, writes
11. Message : "HALTED RUN COMPLETE"
12. GO TO RUN NO. 3

EXCEPTION CONDITIONS

P.T.O.

If correct end of run is not reached GO TO RUN NO. STOP

Retyped
I.M.I.S.
AUGUST 68.

FIG. 10.5. Sample page of operating instructions.

DOCUMENTATION IN THE SYSTEMS DEPARTMENT

I.M.I.S. DOCUMENTATION BRIEF

ASSIGNMENT TITLE	SALES STATISTICS REPORTING SUITE	REFERENCE No.	2Y/SD31
SYSTEMS ANALYSTS	J. Wale		
SPECIAL INSTRUCTIONS	The documentation of the Systems Definition must be open-ended to facilitate extension as further reports arise.	COMPLETION DATE	10 Oct 70
		SIGNED	HV Smith

STANDARD DOCUMENTS

Documents and Modules required are indicated by ✓	TOTAL DOCUMENTATION	FEASIBILITY REPORT	SYSTEMS DEFINITION	OPERATIONS GUIDE	OPERATING INSTRNS.	USERS MANUAL		
	✓	✗	✓	✓	✓	✗		✗

BASIC MODULES

#	Module								
1	HEADINGS	✓	✓	✓	✓	✓	✓		
2	ASSIGNMENT BRIEF	✓	✓						
3	OBJECTIVES			✓	✓				
4	SYSTEM EVALUATION			✓					
5	GLOSSARY OF TERMS	✓		✓					
6	OUTPUT DEFINITION	✓	✓	✓					
7	INPUT DEFINITION	✓							
8	USER PROCEDURES - FLOWCHARTS						✓		
9	USER PROCEDURES - SUMMARY						✓		
10	USER PROCEDURES - DETAIL								
11	COMPUTER PROCEDURES - FLOWCHARTS	✓	✓	✓	✓				
12	COMPUTER PROCEDURES - SUMMARY	✓	✓	✓	✓				
13	COMPUTER PROCEDURES - DETAIL	✓		✓					
14	FILE DEFINITION	✓	✓	✓	✓				
15	CODE LISTS				✓				
16	JOB CONTROL PROCEDURES	✓				✓			
17	OPERATING INSTRUCTIONS	✓							
18	HARDWARE REQUIREMENTS	✓	✓	✓	✓				
19	SOFTWARE REQUIREMENTS	✓	✓	✓					
20	NEW PROGRAM SPECIFICATIONS	✓		✓					
21	NEW PROGRAM DOCUMENTATION	✓		✓					
22	SYSTEMS TEST DATA	✓		✓					
23	IMPLEMENTATION PLAN	✓	✓	✓					
24	MONITORING & DEVELOPMENT								
25	PROJECT TEAM MINUTES								
26	RUN MATERIALS								

FIG. 10.6. A documentation specification.

SYSTEMS ANALYSIS IN BUSINESS

Operating Instructions

This document should be a simple set of instructions for the computer operator, giving just the necessary facts required to run the job in the computer room. The instructions for routine jobs can be permanently arranged in a loose-leaf binder, but a separate run sheet can be prepared for submission with the instructions to indicate run-time options. Alternatively, instructions may be on standard preprinted sheets, marked up to show details of a particular use (see Fig. 10.5).

Each set of instructions should have a title page to identify the job name and number and provide a contents list. The contents itself should include a separate page for each program in the job suite. These pages must give precise instructions to the operator as shown in Fig. 10.5. Note that each page contains instructions to go to other runs or to terminate the job as appropriate according to conditions arising in the run.

GETTING CONTROL OF DOCUMENTATION

From this brief appraisal of the major documents needed in system development, it will be evident that analysts may be reluctant to produce all this documentation in every case. In some instances the material is more or less duplicated in different documents, and there is the possibility of wasting valuable time and resources in producing material that will not be read. Some projects may be simple tasks which warrant a limited subset of the documentation described above. Attempts to produce the full documentation will give rise only to a disregard of the need for documentation generally.

Standards should therefore be flexible and geared to the requirements of particular jobs. One means for controlling this activity is illustrated in Fig. 10.6 where a documentation specification is shown. The systems manager or senior analyst makes entries on to a preprinted form to specify the level of documentation needed for a particular job. This specification will normally be completed early in the project, perhaps at the assignment or feasibility stage; it indicates the major documents necessary and the individual modules of documentation required to build up these major documents. A separate standards manual defining the content and format of each module should be available.

Although this particular example calls off almost all the possible modules, some modules are made to serve more than one requirement. The example also illustrates how the working files of the systems department can be organized and developed to complete important stages of documentation.

A document of this sort is helpful to the individuals concerned since it removes some of the misunderstandings that can arise about what is needed. The analyst receiving such a document knows that the project will be considered incomplete until all the papers have been completed and approved.

Chapter 11

Installing and Monitoring a System

WHEN a system has been fully designed, documented and authorized, the work really begins in earnest. The implementation of the system then starts, and the attitudes of those concerned with its installation and subsequent operation are brought into focus once more. In many instances a new system is intended to alleviate some of the pressures that exist in the user departments, but during implementation the users may have to maintain their current activities and, at the same time, support vital events in the installation of the new system and its subsequent operation.

At the same time there will be many technical problems concerned with file creation as well as the production and introduction of new documentation and new methods of data collection and preparation. These activities have to be very carefully coordinated, particularly if the system is large or complex, and the analyst's capacity as a project manager is tested. Obviously, this phase of the systems task requires the sort of cooperation and control that comes from project team activities. First of all we will consider the basic elements of a good implementation program:

(1) Programming development and testing;
(2) File creation and validation;
(3) Staff education;
(4) Clerical procedure preparation and acceptance;
(5) Changeover plan;
(6) System monitoring and evaluation.

PROGRAMMING DEVELOPMENT AND TESTING

Some aspects of controlling programming activity have been mentioned in Chapter 9. For the most part this phase of a project concerns the people in the data processing department more than it does the eventual users of the system.

However, it is important to keep in contact with the user during this phase and to maintain his interest. There should be a published schedule for the programming activity, and representatives of the users should meet to discuss progress and to coordinate the provision of any supporting activities.

It is also a good idea to interest the user departments in the specification and development of system test data. This data should be small in volume but should, nevertheless, test all the conditions that are likely to arise in live running, with particular emphasis on data errors. The expected results of the test data must also be compiled, including error lists and other printed results provided by the system.

The analyst has overall responsibility for the provision of systems test data, and he must finally decide the conditions that must be tested. To a large extent the provision of test data is a specialist function, requiring a knowledge of the supporting clerical procedures, the computer processing and file maintenance routines, and operating conditions in the computer room. Whereas the analyst cannot delegate this responsibility to the users, it is important to get them aware of the conditions that will arise and to reinforce their understanding of the role that they will play in monitoring and controlling the live system.

Test data must be created in the media in which the programs expect to find it, e.g. punched cards, paper tape, magnetic tape, etc. The analyst may have to arrange that special programs are written to create test files. There is a tendency to try to minimize the difficulties encountered in creating test data by using live files that are already in existence. The danger here is that live data for a given period will not contain all the errors and conditions being tested, and, in any case, much of it may simply test the same conditions over and over again, thus wasting computer time to no great purpose.

FILE CREATION AND VALIDATION

If a system calls for establishing files which have never previously existed as an entity in the organization, then it is likely that a lengthy schedule of events may be needed to develop and create the file. Data may have to be collected manually and be fed to the system – this is often a formidable task, and in an atmosphere in which there may be many live transactions being generated it is difficult to fix accurately the true status of records in the file.

In more favourable circumstances there may already be an existing file which will form the backbone of the new one, and if this file is on some media such as punched cards, it is likely that the initial file can be created by writing one-off programs to edit and convert the data to its final format. In the case of a file which has existed as manual entries on record sheets, it may be feasible to organize a large data preparation exercise to initially create the file from these records.

INSTALLING AND MONITORING A SYSTEM

Most systems entail the collection of transactions which are used to update main files; transactions may be additions, deletions, or amendments to records in the file. It is often possible to create a new file by treating every record as an addition to the file when it is first set up. In this case it will be possible to use the suite of updating programs to create the file, and *ad hoc* programming effort is avoided.

If the file is to be a very large one it may be sensible to phase the creation of it over a long period of time according to some plan agreed with the users of the file. For example, in an inventory control system, certain ranges of items may be brought under control of a new system, leaving the existing system to cope with the remaining stock until all items have been taken on. A personnel records system or a payroll system could be progressively extended to cover the different departments of an organization in the same fashion.

A schedule of this sort has to be known and accepted by the user departments in order that new records entering the file can be adequately maintained as soon as they are taken on. In a file in which there is a high volume of traffic, this can cause many problems to the user who will have to operate two procedures – one to maintain the old files and one to maintain the new.

In some cases, existing records may be so poor that it is necessary to establish an accurate file progressively according to some prearranged plan. In a very large inventory control system a file system could be set up to control goods received and issued without necessarily having absolutely accurate figures for existing stock levels. The stock levels could be checked and transactions frozen to permit accurate reconciliation when a new consignment is received. It might be possible, also, to relate stock checking to some reorganization of the warehouse so that bin locations are audited and handed over to control of the system progressively.

File creation is a problem that concerns the users: they must be prepared for the problems that arise and be able to appreciate the significance of these problems to the development of an efficient control system.

In many cases a computer file replaces several files previously maintained independently on some other media, e.g. an integrated personnel system might replace a payroll file, a personnel records file, a pension fund file, company training records, and so on. The files replaced may have been maintained with different degrees of rigour and on different time bases. Although, overall, one might expect the payroll to be the most accurate file, it is highly likely that in respect of certain data elements other files will be more accurate. The problem is to decide which files are most accurate in respect of which items of data. Where there is difficulty in drawing clear distinctions, programs can be written to compare the files and logically isolate those records which are in conflict.

Duplicate records in manual files, or even punched-card files, are quite common, and these must be identified and corrected by editing and validation runs when file data is taken on. Records which do not match as expected with records in other files must also be isolated for scrutiny and correction. The commonest problem encountered in these circumstances is that keys have been inadequately maintained in some files.

File creation and validation can be a long and arduous task: certainly it requires good relations with those who maintain existing files, whether manual files or otherwise, and a good deal of the work involved may require a tactical approach to meet unexpected data conditions.

Data management software facilitating the preparation of *ad hoc* transcription, editing, validation and updating programs is an invaluable asset to a systems department in file creation. It will enable much of this tactical work to be carried out without resorting to large-scale programming activity; parameters instead can be prepared for general purpose programs to do this work.

STAFF EDUCATION

A most important stage in preparation for a new system is the education of users to support, maintain, and use the system. The education and training plans should include:

(a) Identification of the categories and numbers of staff to be trained;
(b) The time in which training is to be completed;
(c) The types of training needed.

In designing a system the analyst will have identified the categories of staff involved and this will include:

Executive management;
Line management;
Clerical and operational staff;
Data control and preparation personnel;
Computer operations staff.

All who will be concerned in the working of a new system, or who will be concerned with setting objectives for the system, or evaluating its performance, should be included in the training plan.

The training schedule must be planned by the analyst so that users are brought to their peak just as the various aspects go live; clerical staff may otherwise tend to forget what has been learned of the new system. It is important to create enthusiasm for the new project if success is to be achieved, and the line management should be involved in establishing the training program and in presenting it to their staff. It may be possible to link the introduction of the new procedures with improvements in the working conditions and environment of the staff concerned.

Full-time training which requires a lengthy interruption of normal activity is best held away from the usual place of work. The pressures and distractions of the job will otherwise break the continuity of the training, particularly for supervisors and managers. Residential courses are favourable where the training course must span several days or weeks.

Part-time courses sandwiched in between normal working periods are suitable only if the knowledge and skills to be learnt are already very similar to those being exercised by the staff concerned. Training is a subject worthy of fullest management support; if the training program is not carefully planned and executed, one can hardly blame those who operate the system for failing to appreciate the importance of their new role. Computer systems often falter due to relatively trivial incidents arising in the supporting procedures, and inept training plans are often the cause of such incidents.

Training programs may start with formal lectures which provide background to the whole new system, but it is unlikely that these will be sufficiently instructive to obviate more specific and detailed training. General background training is, nevertheless, an important aspect – individuals must be given an appreciation of why the organization is introducing new methods of working, and thereby be helped to appreciate their particular role in the new scheme.

It is generally unwise to have large audiences of different staff categories attending the same lectures. Even those lectures designed to form a background picture have to be pitched at the right level for the audience, and trainees must be encouraged to participate in discussions aimed at their level. It would be wrong, for example, to allow a supervisor in an accounts department to dominate discussions which are primarily intended for delivery van drivers.

Large audiences should be avoided whenever possible. It is usually best to instruct five or ten people at a time whenever dealing with detailed matters; and the instruction should be conducted to allow all who are present to participate intelligently. The training of operatives is best accomplished under the instruction of their management or supervisors, with the systems staff helping out where necessary. The training program could be spread over several weeks: first-line management are briefed; they then brief their supervisors who in turn brief small groups of operatives.

On-the-job training is sometimes possible, but often involves people in performing two tasks – running the new system and the old side by side. Perhaps the best method is to simulate working conditions as they will exist eventually, preferably at some station adjacent to the current operation. Single operatives or small groups can then be taken out of their normal duty for training in the simulated environment. Simulated conditions are also a very useful way of training line management in the use and exploitation of data from a control system.

CLERICAL PROCEDURE PREPARATION AND ACCEPTANCE

The computer routines that form part of a system can be prepared and tested thoroughly before there is a need to use them in live situations. Thus it is reasonable to expect that all major errors in programs will be eradicated before the system is committed to operate on a routine basis. Unfortunately, one often finds that systems do not immediately reach full operational efficiency, and the reasons are often attributable to the interface between the computer and the external clerical procedures which support it.

The preparation for a system should therefore include tests to check the operation of these manual procedures. Such tests may include the completion and handling of source documents, data control checks, data preparation, batching and control of input, dealing with error reports, and use of computer output reports.

The education of users previously mentioned goes a long way towards preparing for new procedures, but often the true impact on users may not be felt until the work is actually done. This underlines the need to simulate the live environment in training schedules. The keynote really is to be satisfied that the users are aware of the full implications of a procedure, and that they understand and accept the activities with which they are involved and the media with which they are to work.

Input

The format of input documents is determined at the system design stage, and will include agreement to:

(a) Number of copies;
(b) Their distribution;
(c) Colour of copies;
(d) Size;
(e) Suitability of format.

Those who are to handle the documents should be given the opportunity to test out their efficiency in some pilot scheme before large volumes of the documents are ordered for the live system. The whole procedure from origination of transactions, through data collection, to computer input should be tested in this way. This will help the user to appreciate the task and perhaps allow improvements to be made to the original proposals before implementation.

Procedures for control of input data should be affirmed at the same time, and the design and layout of control sheets for recording control figures should be put to the test at the same time.

Error Control

Correction of errors is a major clerical function, and if the relevant procedures are not properly evaluated or understood, the value of the system concerned may be seriously undermined. Validation routines are potentially the biggest source of error reports, but updating programs may produce more enigmatic conditions.

Error reports from validation routines are usually based upon logical analysis of data elements within particular transactions; some diagnostic criteria can often be derived from context and included in the error listing. With updating programs errors may arise because keys cannot be matched; this may infer that a main file record is missing or has had its keys amended by a previous transaction, has been incorrectly recorded with the wrong keys, or perhaps the transaction is incorrectly coded.

In the simplest circumstances the data control clerk dealing with these discrepancies will look up the original transaction document to see if it has been recorded correctly, and this may entail contacting the originator of the document to check his records.

Some errors arising in an updating system may result from the accumulative effects of several previous transactions, e.g. a computer file may record a stock quantity for a product when there are clearly none in stores. This type of error may require an audit trail through previous weeks or months of work to establish the correct situation.

The problem of assessing this work is that the amount of effort expended on individual queries varies considerably, and also it is not easy to assess the overall volume of error lists before the system has had a chance to run in a live environment. Nevertheless, attempts must be made to estimate volumes and the nature of errors that arise; these estimates should be discussed with the supervisor or manager who will be responsible for corrective action. The format and contents of error reports should be subject to close appraisal before final approval.

In the early stages it pays to budget for effort in excess of that anticipated so that the file system has every chance of getting through the difficult teething period. If a file system is allowed to absorb a lot of unactioned errors, the use of the file for any functional purpose may be severely limited.

Use of Output

The comments applied to error reports above are also relevant to the use of any output – whether it is a document providing general information or one used for a specific activity. A system should never be allowed to go operational without simulating the use of the output to test the relevance of its content and format to the activity concerned.

Where a system deals with entirely new data conditions that are difficult to simulate, then a phased implementation plan is often necessary to test and install features of the system progressively, perhaps dealing with certain departments of the organization in turn until the whole is operational.

Acceptance of System Timings and Loads

Often, the measurement of supporting clerical activities is regrettably overlooked, and systems are frequently forced to 'go live' in an atmosphere of crisis. A typical condition is that there are too few resources to start the new system and run the existing system.

The symptoms are usually the inability to get on top of data errors and delays in system schedules which have harmful effects upon operating conditions in the departments concerned. Attempts to catch up inevitably cause strain among the user department's staff, and a fundamental disregard for the system can set in before it has a chance to show its advantages.

The analyst has the task of estimating the loads generated by his new system both during normal running and in the difficult changeover period. The response required from certain departments and individuals has to be assessed and published, and the approval of the user authority should be obtained before implementation commences. The analyst may have to be quite forceful in his approach to the management at this stage in order to be satisfied that they appreciate the nature and volume of work to be handled during implementation and normal running.

CHANGEOVER PLAN

A system *goes live* when it starts to process real data and to control real events rather than simulated conditions. To successfully complete this phase of a job, the analyst must have prepared the ground thoroughly as described in the preceding sections of this chapter; but the moment of truth seldom arrives with every aspect of the plan going to schedule. The most intangible problems include the readiness of the users to provide the supporting functions, and perhaps the quality of the data initially accepted into the new system.

A well-devised education programme and a well-executed file creation exercise should have minimized these problems, but the analyst has to be poised at this time to deal with outbreaks of trouble at various points in the system. The changeover to the new system should be conducted, as far as possible, to minimize the number of trouble spots that can arise, e.g. implementation should be phased rather than attempted with a 'take off or bust' outlook. Events and circumstances may have some compelling logic of their own which determine the mode of changeover, and some of these constraints are considered below.

Firstly, there is the problem of deadlines which may have been agreed as important objectives for the company to achieve at particular dates. If the system is instrumental in meeting such objectives, then every effort is needed to launch it on time. In these circumstances it is preferable to set up a working party of users and systems staff to supervise the introduction of the system so that problems are identified and tackled as soon as they become troublesome. Delays and achievements in implementation should be discussed and published so that all those relying on the new system can make necessary arrangements to secure extra resources or back-up facilities.

If the movement of transactions is very dynamic, the new system may be obliged to accept the full load immediately upon going live. Perhaps the load is such that it is impracticable to run the old and new system side by side. Sometimes the continuity of the data is such that data collection systems for the old and new systems are incompatible.

In almost every case, however, the changeover should be conducted to reconcile the closing position of the old system with the opening position of the new system. This is obviously a necessity for systems which fulfil some statutory requirement, but is just as important for any system which is used to generate management information or to directly control some process.

The introduction of new procedures should not be timed to coincide with peak loading conditions, perhaps attributable to other systems and functions, in the departments concerned.

Pilot Running

Pilot running entails selecting a small part of the total organization for the live running and proving of a new system, e.g. one sales area may be chosen to implement a new order processing system, or one stores depot to run live a new stores control system.

The essential reason for selecting a particular area must be that it will provide an exacting test of the various facets of the whole system. As the system is proved and debugged entirely, so it can be extended to other areas.

Parallel Running

Parallel running means testing the new and the old system side by side using the same data and comparing the output produced by both systems until the new one is proved to be producing the desired results. This always presupposes that the old system is working correctly – in practice one often finds that introducing a new system, and thereby processing the data in a different way, merely uncovers more inaccuracies in the old system. It is not always easy to make comparisons between two systems because their characteristics may be different; but there must always be some objective established for the new system and, where this can be expressed in quantitative terms, the systems staff and the users should be able to establish whether these aims are being met.

SYSTEM MONITORING AND EVALUATION

System monitoring must take place continuously throughout the life of a running system so that if it begins to fall behind on schedules or fail to meet its objectives for quality or relevance of output, a more detailed investigation can be started to determine whether it is a candidate for further development.

Changes will probably be enforced at some time or another because the environment and structure of the organization itself changes.

Diversification of products, fiercer market competition, different methods of manufacture, mergers with other organizations, and so on; all these changes will strain even the best and well-drilled procedures. Changes in the outlook or ambitions of management can affect the relevance or effectiveness of a system. Perhaps, more than anything else, the continual need for more accurate and timely information will imply change.

Hopefully the analyst will have built flexibility into his system; particularly he should have allowed for increased volumes of work. Additionally he should have allowed for ease in changing input formats, file formats, output reports and processing routines. Code structures should also have been designed with future expansion in mind.

The systems department should review all systems with the relevant managers at regular intervals – perhaps every week during the first months of operation, and never less frequently than once a year. The evaluation of a system has to be made in relation to the current objectives of the organization For example, the service afforded by an order processing system may be inadequate simply because two years after its installation a much higher service is needed. The original aims must first be compared with current expectations to see whether they are in line, and, next, performance levels for the future should be set for further comparison. If it becomes clear that a system is failing to produce the results needed, then an assignment should be raised to study the nature of the problem and perhaps a detailed feasibility study will then be called for. The manifestation of system weaknesses include the following symptoms:

(a) Poor adherence to processing schedules;
(b) Forms and data collection media containing comments and *ad hoc* remarks to cater for new situations;
(c) Lengthy error lists from validation and updating programs;

INSTALLING AND MONITORING A SYSTEM

(d) The inability to reconcile the results with results of other adjacent systems;
(e) Operatives or machines being idle awaiting output;
(f) Output reports remaining unread or not questioned by users;
(g) Excessive overtime being regularly performed to keep work flowing;
(h) Complaints or loss of business due to customer dissatisfaction with service;
(i) Relatively high stock levels in relation to demands.

These are the things the analyst should look for and be prepared to react to. A system is never complete, and at any time can be regarded as at a stage in a process of evolution towards an ideal condition. The important thing is to aim each stage of development at specific objectives, and then go for them vigorously. Each advance will inevitably create potential for future progress, and should permit the eventual enhancement and integration of the organization's total procedures.

Chapter 12

Introducing Change

INDUSTRIAL nations are destined for continual change if they are to maintain their standards and expectations for secure and peaceful existence. As the process of industrialization spreads, the successful nations are those who react most quickly to technological changes, and to change and growth in the world markets. Information handling systems are the key to successful operations, and are increasingly used to control industrial and commercial policy and to provide control systems for managing both government and business agencies. Future computer applications are likely to require many interconnecting streams of information which will cut across present functional and organization boundaries and will affect all walks of life.

Hospitals, local government departments, state-owned corporations, industrial and commercial organizations are all using computers in such a way that the man in the street is feeling the impact. Some of the changes brought about sometimes appear to be detrimental to individual social interests, and may engender resistance from the parties involved.

As an agent for change, the systems man soon becomes aware of reaction to his attempts to systematize things. Both the layman and the well-informed individual may be predisposed to resist the planning and introduction of computer systems, and the systems analyst has to identify the reasons and make sincere attempts to resolve conflicts. In this chapter some of these problems are discussed.

SOURCES OF RESISTANCE

A systems analyst has many contradictory calls placed upon him in his everyday life – he is a change-maker and has to have his finger on the pulse of the organization. The aims and ambitions of the individuals within the company are often directly affected by systems development activity, and this may engender a resistance to projects and cause attempts to divert projects into channels that suit specific needs and not others.

INTRODUCING CHANGE

A mature analyst often has to play the role of a manager, since he has control over the utilization of costly resources. However, this control is often exercised by proxy – the systems analyst has no direct authority within the user organization and extends his influence by being able to encourage managers and supervisors to develop, accept and install new working methods which will contribute towards the successful implementation of his assignment.

The systems analyst has to spend a great deal of his time studying the attitudes and needs of the people running the current system, and must seek to gain their confidence and support before undertaking any significant step. There may be a number of independent groups concerned with a particular project, and their aims may be separate and sometimes contradictory. Some problems may arise from a weak organization structure within the company, and these aspects are almost certain to present themselves as potential stumbling-blocks in the creation and initiation of a new system.

Changes in working methods are aimed at eliminating the duplication of functions in different areas of the company and to provide streamlined administrative systems that are able to respond to the changing business environment. The new procedures must themselves be flexible, otherwise they will quickly fall into disrepute. The creation of such procedures inevitably disturbs the stability of an organization. Managers and supervisors within the problem area may feel that this change implies a criticism of their own operations. This may engender resentment which will be enhanced if the persons concerned have been working hard at the old system to keep up with the work.

New working methods are particularly sensitive in production and distribution organizations, and full discussions should take place in joint consultation with workers' representatives to attempt to remove fears of redundancy, loss of status, or reductions in earnings. To a large extent the analyst is governed by the policy and attitudes within his company relating to industrial consultation. If the company has a history of fair dealing in previous situations, then the analyst has a good start. In any event all proposals of this nature should be discussed initially with the industrial relations department, where one exists, and with chief executives in the division under scrutiny. All such problems must be identified by evaluating the effects of new proposals upon the individuals concerned.

Each member of an organization has some understanding of his own role in life, of his relationship with those around him, and of the aims of his department and company. This understanding is obviously related to the individual's background, previous learning and the present nature of his work.

Everyone in a business organization expects payment for their services, but, in addition, most of us seek self-respect in the work that we do. Security is related to the payment received, but also the nature of the work done – interest in the job and in the surrounding social atmosphere is often just as important.

Resistance can arise from the mere fact that change is being considered purely because the social environment may be threatened. This is particularly evident where there is a need for rationalizing the geographic distribution of units in an organization. Here there are also powerful influences outside work both in the home and other social organizations. If an analyst is working against this background he cannot expect to succeed if the company concerned has not adopted a creative policy, but success is also related to the analyst's ability to understand the policy and the position of the individuals. In this respect the analyst is the man in the middle, and he is often the focal point at which the aims of the company and its members meet.

In matters involving industrial relations policies, the analyst has to be well informed yet be discreet. He should do nothing to embarrass either party when consultation is taking place, and should always ensure that the machinery for joint consultation is given time to operate before he allows matters affecting working conditions to be given broad publicity.

However, the analyst is not a specialist in industrial relations and, indeed, provided he understands the nature of such problems and behaves in a prudent manner, one should not expect him to accept any great responsibility in these affairs.

THE USER'S VIEW OF THE SYSTEM

A large company can be viewed as having a certain personality which manifests itself in outside contacts with other organizations and in attitudes existing between individuals within the company concerned. In fact this personality is compounded of a number of attributes, aims and ambitions of those individuals that make up the company. The extent to which these can be placed into a formal set of rules depends largely upon the nature and development of the organization.

When an analyst is at work on a business problem he is generally working within a framework which has been set for him in his assignment brief, but may also have to observe matters of general company policy, e.g. as reflected in company instructions which are broad directives in how to act in particular situations.

It would seem, therefore, that the analyst should have clear guidelines in how to conduct his assignment – but in practice such guidelines cover only the broadest policies and are often open to interpretation. It is the management within the organization that determines its mode of operation; from the directors down to the supervisory level, discretion is exercised within certain bounds. Since a system exists to support the managers in achieving their aims at their particular level; one must expect pressure to arise to direct the project into certain channels. The analyst is often in the middle of controversy arising from different expectations of the executives or line managers concerned with a project.

It is for this reason that computer projects have to be supported at the highest level in the organization. The principles laid down in Chapter 9 concerning the selection and development of projects are extremely important in avoiding friction arising from a misunderstanding of the aims of an assignment. Difficulties may still be encountered in the detailed design of procedures, but these will tend to be about ways and means of securing particular objectives and, as such, the contending views should be capable of evaluation.

Where there is a very strong reaction between one manager and another the analyst must take care not to appear to take sides; if particular issues are unnecessarily obstructive to the development of a task, then they must be referred to the executives who authorized the original study. This action is usually best accomplished by the systems manager; and it is to be hoped that this action is necessary only in extreme cases. It is a bad sign if an analyst has frequently to resolve issues by asking for such pressure to be applied. Major issues should be conducted openly and rationally within project teams; in this way the analyst can avoid the appearance of bias towards individuals.

A sincere attitude to all problems is an essential characteristic for the analyst. When unpopular courses of action have to be recommended in reports, it comes much easier to the affected parties if they are able to see that a fair and rational appraisal has been made, and particularly if they have participated in the appraisal. If the analyst has created an atmosphere of intrigue or has shown impatience or intolerance when issues are discussed, he will not get support from the users, and the project will suffer.

It is important that the systems development group should be seen to conduct their work in an impartial manner in pursuit of known and publicized aims. Where suspicion is harboured about the intentions of a project, resistance will be forthcoming. This resistance can arise both from line management and the operatives responsible for particular functions.

In the case of the operatives they will principally be concerned about their future security, status and conditions of employment. Factors affecting such conditions should be identified as soon as possible and discussions should take place to see that management are aware of the possible effects and are able to formulate plans to meet them.

Resistance from management can arise for a number of reasons, but the following may be motives for resistance:

(a) General distrust of computers;

(b) Fear of the intentions of higher management;
(c) Distrust or disregard for the systems department;
(d) Apprehension about changes in the stability of their particular department;
(e) Possible effects of subsequent changes in organization or responsibility.

The analyst has to bear these fears in mind and try to analyse the effects upon individuals or groups. Wherever possible the systems man must conduct his work so that he is able to diminish these fears if they are imagined. Nevertheless, some fears may be relevant to the task being undertaken, and therefore the systems group and the senior executives concerned should be prepared to take steps that will preserve the security and career prospects of the individuals concerned.

Selling the need for change and stressing its benefits to the organization as a whole is vital to gaining acceptance of a new system. Even where good attempts are made to sell new proposals, the fears themselves can never be entirely resolved while design work is being conducted. The analyst will find himself continually bumping into potential conflicts as the detailed design continues. True, these conflicts may be at a relatively low level once the overall outline of the system has been perceived and agreed, but, nevertheless, these conflicts are just as damaging to the success of any project. Failure to gain a willingness to adopt new working methods can disrupt any system.

Whenever conflicts arise, the analyst has to be tactful in preventing issues from boiling over until such time as he is able to report back to the relevant managers so that a sincere attempt can be made to resolve the difficulty.

AUTOMATION PHOBIA

Most people will have some preconceived notions of what a computer does. On the one hand, this may be an abject fear of the de-humanizing potential of machines or an over-optimistic view of the immediate potential of a computer. General education about the nature and applications of computers can help to alleviate these fears; in the long run it is participation in systems development and implementation that secures more enlightened attitudes. It soon becomes apparent that many benefits can be achieved by using computer power to enhance the potential of particular individuals or groups, but that these advantages are not easily won without a considerable change in working methods. If a system is developed properly, these changes should release individuals from drudgery to a more rewarding role in which control and discretion can be more readily exercised.

INTRODUCING CHANGE

FEARS OF HIGHER MANAGEMENT

There can be no doubt that when a computer is used correctly it can satisfy operational requirements and at the same time provide much information in a summarized form to higher management. The use of computers in this way enables management to monitor and observe the performance of others further down the line. If a person or a group have qualms about their own performance or interpretations that may be derived from the data, there will inevitably be a reaction when developing, implementing, and running the system.

However, systems can be designed to also highlight genuine problems that particular managers or groups have to contend with, and can point to improvements that are necessary to assist in fulfilling their particular function. If a system is intelligently designed, the users will feel that it enables them to understand their own operation and to provide facts which describe the environment in which they are operating.

The systems department often forms part of the central management services group within a company, or perhaps in smaller organizations may report to the chief accountant. It is easy for users to view the systems department as part of a corporate tentacle stretching out into the separate operating departments and divisions in order to gain centralized control. To some extent this may be an aim of the whole data processing operation, but, if so, it is important for the divisional management to feel that they have control over the selection and direction of projects so that they are able to gain effective control of their own operation. This gets us back once more to the factors stressed in Chapter 9 regarding the selection and control of projects.

The user's attitude may be affected by his apprehensions concerning the security of data in files maintained for him. It is a negation of the principles of information systems to make strict rules of general application about who may receive what information, but there is clear justification for identifying certain types of data as confidential. This does not mean in every case that the data is absolutely for exclusive use; in many circumstances it is merely necessary to ensure that the users have reached a clear understanding before a particular use is made of a file. The systems group will have to satisfy users that adequate security procedures can be applied to their files, and this should be reflected in the whole attitude and approach of the systems staff.

USERS' VIEW OF THE SYSTEMS ANALYSTS

The line management must have respect for the role and abilities of the systems department if systems projects are to reach fulfilment. One of the most important problems is to be able to assist line managers by responding quickly to critical situations that arise. Most major systems projects have long lead times, and the users often misunderstand why this should be so. Even if they understand some of the problems they cannot be expected to simply sit back and wait while a system emerges. The users may have problems which have to be encountered as they arise with whatever resources are at hand; and the systems group should expect to provide whatever tactical support they can.

Such support may include setting up simple files and reporting systems to provide temporary facilities for key problems. If these projects are sensibly conceived and implemented, they can be tackled as steps in an overall plan and will dovetail into the main-line development function which can be proceeding in parallel.

In this way the users can be helped to realize that the systems group is not just for egg-heads absorbed in abstruse and complex data processing problems but that they are capable of effective and worthwhile action to meet both short- and long-term problems.

The systems analysts should be trained in general management principles and techniques and given a thorough grounding in the aims and objectives of the particular company and its integral departments. In this way they can be equipped to talk realistically about users' problems and methods for solving them. The use of a computer then becomes an incidental factor in the relationships between the analyst and the user.

The ideal relationship arises where the executives of a particular division regard the systems analysts as an essential part of their management team. The analyst thereby becomes a troubleshooter who has a dynamic role in applying data processing technology to their problems. He helps to identify, describe, and resolve problems with management.

EFFECTS OF CHANGE UPON DEPARTMENTAL STABILITY

If a department has had for several years a relatively stable existence, the staff will be familiar with the existing system and will know its relative strengths and weaknesses. Thus they may well have the ability to get results despite day-to-day operating difficulties. Attempts to introduce new methods may destroy some of the basic familiar disciplines upon which the user's staff are dependent, and therefore resistance will arise consciously or otherwise to the new methods.

To overcome this resistance, a carefully prepared education programme has to be developed to indoctrinate the staff concerned. They will want to know what the effect of the new methods will be on their working conditions, particularly where there is a significant change in the responsibilities of individuals or in organization structures. It is also useful to introduce change gradually, making sure that one aspect of the system is implemented correctly before introducing the next.

The education plan should explain the reasons for introducing new methods, so that all members of the organization share in the overall aims and objectives of the system. New methods should never be forced upon a department who are not ready for them. If a system gets off to a bad start it may be very difficult to maintain morale and confidence, and a hurried compromise may be reached to keep the system running.

Where there are a large number of people affected by a new system, it is necessary to ensure that seeds are sown in a controlled manner. It is not good enough just to issue a new procedure manual to everyone concerned. Neither is it satisfactory to make mass presentations to large numbers of people at any one time; groups of 5–10 people provide a much better audience for presenting new proposals. In such presentations there should be plenty of opportunity for discussion that will serve to identify the fears or reservations of the user group. The presentations can be made by the system analysts, if necessary, but it is a good idea for managers and supervisors to take a leading role when presentations are made to their staff.

Perhaps the most potentially damaging problems for a new system are those that arise where departmental responsibilities are at stake. It is often necessary to so reorganize a company that certain managers and their staff may become redundant to the particular operation. Particularly does this arise when a new system is being developed to centralize control in a large organization which may have several operating divisions. Accounting and statistical reporting functions which may have been duplicated in each division will perhaps be run-down to provide only for sufficient local control and co-ordination with the central system. The decision to embark upon such a course will inevitably have been generated by top management, and the systems analyst is thereby acting as an instrument for implementing this policy. He will perhaps have the task of setting up a system which will need much local support in its early stages but perhaps a much lower level of activity later on.

Here the systems department are once again batting on a sticky wicket if the senior management have not resolved how to tackle the ensuing reorganization. Individuals affected must be given opportunities elsewhere before the full impact of the changeover is felt.

Apart from the human problems involved, there is also the need to convince local management that the new system will provide improved control and support for their own activities.

In business organizations it is often found that the people most affected by new procedures are those concerned with clerical functions. In some cases redundancies ensue immediately from a new system, but invariably the new procedures bring with them more exacting standards for data collection and control. Error reports and exception conditions generated by the system have to be dealt with, and sometimes work has to be turned around much more quickly than before in order to comply with computer room schedules. In practice, therefore, it is the nature of the work that changes and staff reductions, where they are required, can be effected over a fairly lengthy period of time in which the development is phased.

The changes in working conditions must be studied to ascertain the effects upon individuals and the way in which they fulfil their particular task. Too often costly computer procedures are implemented without sufficient regard to the impact upon the volume and quality of work required from supporting staff.

THE FUTURE CHALLENGE

In the last resort it is the systems analyst who is responsible for designing and getting a system running, and, in so doing, he is continually concerned with changing attitudes and habits of people.

The social problems of implementing computer systems appear largely to be matters of common sense, yet they are well worthy of attention. One can observe so many systems which fail to mature or which do so only after excessive delays. Some grounding in human relations is an important asset for any systems man, and the more basic works of industrial psychologists should be part of any syllabus for training analysts.

The efficiency of any enterprise is very much dependent upon good organization and methods of communication, and these can only be achieved in an atmosphere conducive to cooperation. Systems should be designed to help people and to support them in their working lives. Both the operative and the decision-maker need to view computers as providing them with a stable background in which to face today's problems and to meet changes required in the future.

Perhaps the future success and prosperity of our industrial societies depends not so much upon the ability to gain technical mastery over computers, but upon the power to implement systems which provide a secure social environment for individuals and yet give our industries streamlined and efficient control systems.

Bibliography

HARDWARE AND FUNDAMENTALS OF COMPUTING

1. Herbert Maisel: *Introduction to Electronic Digital Computers* (McGraw-Hill, 1969).
2. Anthony Chandor, John Graham and Robin Williamson: *Dictionary of Computers* (Penguin, 1970).
3. Gerald R. Peterson: *Basic Analog Computation* (Collier–Macmillan, 1967).
4. British Computer Society: *Character Recognition* (1967).
5. British Computer Society: *Data Transmission Handbook* (1964).
6. S. H. Hollingdale and G. C. Toothill: *Electronic Computers* (Penguin, 1956).

PROGRAMMING

7. R. Clay Sprowls: *Introduction to PL/1 Programming* (Harper & Row, 1969).
8. D. D. McCracken: *A Guide to Fortran Programming* (Wiley, 1969).
9. D. D. McCracken: *A Guide to Cobol Programming* (Wiley, 1963).

O. & M. AND METHODS STUDY

10. Victor Lazzaro (Ed.): *Systems and Procedures* (Prentice-Hall, 1958).
11. H. P. Cemach: *Work Study in the Office* (McLaren & Sons, 1969).
12. H. Symes: *Office Procedures and Management* (Heinemann, 1969).

COMPUTER SYSTEMS DESIGN

13. Anthony Chandor, John Graham and Robin Williamson: *Practical Systems Analysis* (Rupert Hart-Davis, 1970).
14. H. D. Clifton: *Systems Analysis for Business Data Processing* (Business Books, 1969).
15. J. Martin: *Design of Real Time Systems* (Prentice-Hall, 1967).

HUMAN RELATIONS

16. Mrs. E. Mumford and T. B. Ward: *Computers – Planning for People* (Batsford, 1968).

17. Abraham J. Siegal: *The Impact of Computers on Collective Bargaining* (M.I.T. Press, 1970).

MANAGEMENT INFORMATION

18. T. R. Prince: *Information Systems for Management Planning and Control* (Irwin, 1966).
19. H. L. DAVID: *Accounting Computers Management Information Systems* (McGraw-Hill, 1968).

Index

Acceptance procedure   154–5, 190
Address generation   134
Analysis of existing procedures   47–51
Analyst
  attributes of   22–6
  role of   12
  technical qualifications   27–8
Approval
  of output   64
  stages of system   149
Assessment of costs and benefits   56–9
Assignment brief   35, 151–2, 169
Auditors
  liaison with   72–3, 169
  system controls   72

Balance, input/output   120–1
Batch controls   72–3
Batch processing   103
Block diagrams   46
Blocks
  and file organization   96–7
  fixed-length   97–8
  variable-length   99
Buckets   129
  capacity   130
  logical numbering of   130–1
  overflow   135
  packing density   131

Cards   *see* Magnetic cards; Punched cards
Changeover

  planning of   190
  scheduling of   190
  types of   191–2
Character recognition   83–90
  mark reading   87–9
  MICR   83–90
  OCR   85–9
Charts   *see* Flowcharts
Checking
  for errors   104–5
  for parity   102
  for programs   168
Classification
  of data elements   110–11
  of facts   47–50
Codes
  paper tape   81
  punched cards   76–9
Coding of programs   18
Collection of data   74–92
Communication in system development   169
Compilers   11
Computer operations   20–1
Control
  in operation of systems   70
  in system development   147–68
  of programming   161–8
Control accounts   72
Conversion of existing files   184
Core storage, utilization of   120
Costs
  in operation of systems   56–9, 174
  in system development   59, 173

205

INDEX

Creation of files 134, 184

Data
  for program testing 168
  for system testing 167
Data characteristics 67
Data collection 66–8, 74–92
Data control 27
Data elements 110–11
Data management 163
Data organization
  for direct access devices 129–31
  for magnetic tape 111
Data preparation 76–81
  magnetic tape 80–1
  paper tape 81–3
  punched cards 76–81
Data processing 14–22
Definition
  of problems 35, 151–2, 169
  of systems 177–9
Design
  of clerical procedures 68, 87–91
  of data collection 68, 74–91
  of files 69
  of programs 163
  of random files 133
  of sequential files 132
  of serial files 132
Digital computers 11
Direct access files 126–40
  equipment 126–8
  file structures 126–35
  indexing methods 134
  processing methods 127–8
  storage principles 128–31
Discs 127, 129
Documentation 169–82
  assignment brief 35, 169
  existing procedures 38
  feasibility report 152, 154, 171–5
  operating instructions 179–82
  operations guide 179
  programming specifications 166–8
  systems definition 155, 157, 177–9
  working files 175–7
Dummy reports 63
Dumping and restarting 118

Editing 105

Education
  of systems staff 26–8
  of users 186–8
  planning for 187
Elapsed time
  of peripheral operations 122–4
  of progress 124
Error control 189
Error files 112
Error reporting 116–17
Exception reporting 64–5

Fact recording 40–7
Fail safe 146
Feasibility approval 154
Feasibility reports 60, 154, 171–5
Feasibility studies 52–60, 152–3
File
  creation 184–6
  descriptions 47
  design 69
  documentation 47
  maintenance 108–16
  mapping 130–1
  packing density 131
  processing methods 127–8
  security and protection 102
  structures 132–3
  re-organization, 137
  updating 108–16
File types
  random 133–4
  sequential 132, 134
  serial 132
Fixed- and variable-length records 97
Flexibility 63, 162–3
Flowcharting methods 44–7
Flowcharting symbols 45
Flowcharts
  computer runs 46, 47
  detailed program 17, 166–7
  outline program 16, 166
  procedure charts 44
Form descriptions 42, 43
Formats, approval of output 64
Functional relationships 33–5

Hardware controls 70

INDEX

Hardware failure   144
Human relations   27

Identification
  of data elements   66–7
  of input needs   66–7
  of management needs   148
  of objectives   148
  of output requirements   63–4
  of problem boundaries   148
Implementation of systems   183–93
Indexing
  input transactions   106–7
  of sequential files   134
  partial   134
Inhibit rings   103
Initial investigation   30, 148
Input controls   72
Input editing   105
Input methods and media   74–92
Input transcription   104
Input validation   104–5
Input/output balance   120, 121
Interrogation of files   142
Interviewing techniques   39
Investigation
  existing procedures   35–40, 50–1
  purpose of   37, 50–1

Keyboard input   142
Keychanging methods   113–16
Keys
  and sorting   109
  of records   109, 128, 132

Languages
  programming   11, 143–4
  scientific   11, 143–4
Layout of records   131

Magnetic cards   126
Magnetic character recognition   83–90
Magnetic discs   126–9
Magnetic drums   127
Magnetic tape   93–8
Magnetic tape encoders   80
Magnetic tape files
  basic handling methods   93–4

  basic run types   99–104
  charactistics of media   95–6
  records and blocks   96–9
  reporting from   102
  security of data   144–6
  sorting   101
  timing of processing   121–5
  updating files   101, 108–16
  utilization of media   97
  variable- and fixed-length records   97
Management
  identification of requirements   148
  involvement in systems   148–55
  relations with systems staff   169
  steering of systems activity   150
Mapping   130
Mark reading   87
MICR   83–5
Multi-access systems   143
Multiprogramming   *see* Timesharing

Objectives for system development
  61–2, 148
OCR   85–90
On-line input   90–2
On-line working   141–6
Operating instructions   168, 170, 179–82
Operations department   20–2
Operations guide   179
Optical Character Recognition   85–90
Organization
  of data processing   12–20
  of random files,   133, 134–5
  of sequential files   132–3, 134
Organization and methods   27–8
Organization charts   41
Output
  acceptance and approval   64
  design   62–4
  effectiveness   62, 189–90
Overflow
  evaluation of   139–40
  random and sequential files   134–5
  retrieval of records   136–7
  timing of processing   139–40

Packing density   131

207

INDEX

Paper tape 81–2
Parallel running 191
Parity checking 102
Partial indexing 134
Participation
  of managers 148–56
  of users 156–7
Peripheral devices 126–7
Periperal limited programs 122–5
Peripheral times 138–40
Peripheral utilization 130–1
Permit rings 103
Phased implementation 191
Pilot running 191
Planning
  education of users 186–8
  implementation 190–1
  introduction of change 194–202
  system development 158–60
Preparation of data 76–81
Procedure acceptance 188
Processing
  batch processing 126
  random files 128–35
  real time 142, 145–6
  sequential files 128–35
  serial files 128
Processor limited programs 122–5
Program
  design 163
  detailed flowcharts 17
  documentation 166–8
  flexibility, 163
  outline flowcharts 16
  specification 162, 166–8
  testing 19, 165, 183–4
Programming department 15–20, 161
Programming, stages of 16–18, 164–5
Project control methods 147–61
Project teams 153, 156–7
Project terms of reference 33

Random files 133
Real time systems 142–3, 145–6
Recognition of characters 83–90
Records
  disc files 130–1
  fixed-length 97
  magnetic tape files 95–9

  serial files 96–9
  variable-length 98
Re-identification of records 112–14
Reliability in on-line systems 144
Remote terminals 142
Reporting 116–17
Restarts 118
Run types
  direct access files 128
  magnetic tape files 104–11

Scheduling computer runs 118–20
Scheduling programming 163–4
Security of data 102–3
Seek areas 129
Selective/sequential processing 128
Sequential files 128, 132–3, 134
Serial files 128, 132
Serial processing 128, 132
Simulation of operating environment 187–8
Software 140, 162–3
Sorting 107–8
Specification of systems 155
Stages of systems analysis 147, 149
Standards of documentation, 170
Steering committee 148–9
Storage utilization 130–2
Symbols for flowcharting 45
System monitoring 192
System timings 138–40, 190
Systems analyst 12
  attributes of 22–6
  technical qualifications 26–8
Systems controls 70–2
Systems department 12–20
Systems design 14, 21–2, 147, 149
Systems development 21–2
Systems implementation 183–93

Tagging 136–7
Test data
  for procedures 166–7
  for programs 168
Testing
  on line 143–4
  programs 167
  remote 143–4
  systems 167

Timesharing   143
Timing
  direct access files   138–40
  magnetic tape processing   121–5
  of computer procedures   121–5
Transcription programs   104
Turn-around documents   88

Updating
  of direct access files   126–40
  of magnetic tape files   108–16
User attitudes
  to change   196–8
  to computers   198
  to systems analysts   199–200
User participation   156–7

Validation of files   184–5
Variable-length records   98

Work measurement   28
Work study   12
Working files   175–7
Write permit or inhibit rings   103